W9-CAY-860

More praise for

THE WISDOM OF WHORES

A *Library Journal* Best Consumer Health Book of 2008

"Required reading for anyone who works on HIV and AIDS or in the broad field of global public health."
—Richard G. Parker, PhD, *New England Journal of Medicine*

"Elizabeth Pisani combines brains, passion, and a welcome dose of irreverence and mirth [as she] powerfully connects frightening statistics to the real-life stories of a wonderful case of insightful sex workers, smack addicts, transgenders, and other outcasts. . . . She's blunt, bold, and relentlessly honest, even about the limitations of her own efforts, which make her prescriptions all the more convincing."
—Jon Cohen, author of *Shots in the Dark: The Wayward Search for an AIDS Vaccine*

"It's rare to find a great public policy book that also happens to be a page-turner—but *The Wisdom of Whores* pulls it off. This is a must-read for epidemiologists and a great read for everyone else."
—Charles Wheelan, author of *Naked Economics: Undressing the Dismal Science*

"Pisani prefers to hit the controversy head on . . . her impassioned critique of failed prevention programs and distorted aid spending is never dull." —Laura Blue, *Time*

"Pisani's messages are timely." —Andrew Jack, *Financial Times*

"This is an utterly fascinating book. . . . Elizabeth Pisani writes with enormous verve and acerbity, her prose alive with anecdote and metaphor. . . . *The Wisdom of Whores* is a great read." —Stephen Lewis, *Globe and Mail*

"[An] account of spin, waste, and denial." —*The Economist*

"Elizabeth Pisani presents the world of AIDS prevention with humor, attitude, and heart." —*Next Magazine*

"[Pisani] delivers a strong, well-told and believable message." —*Kirkus Reviews*

"A blunt, cynical, and even funny insider's view of global HIV-prevention efforts. When she isn't telling colorful stories, she's skewering everyone who allows ideology to overrule science." —*Library Journal*, starred review

THE WISDOM OF WHORES

THE WISDOM OF WHORES

Bureaucrats, Brothels and the Business of AIDS

ELIZABETH PISANI

W. W. Norton & Company
New York London

Copyright © 2008 by Elizabeth Pisani

First American Edition 2008
First published as a Norton paperback 2009
All rights reserved
Printed in the United States of America

For information about permission to reproduce selections from this book, write to Permissions, W. W. Norton & Company, Inc., 500 Fifth Avenue, New York, NY 10110

For information about special discounts for bulk purchases, please contact W. W. Norton Special Sales at specialsales@wwnorton.com or 800-233-4830

Manufacturing by Courier Westford

Library of Congress Cataloging-in-Publication Data

Pisani, Elizabeth.
The wisdom of whores : bureaucrats, brothels, and the business of AIDS / Elizabeth Pisani. — 1st American ed.
p. ; cm.
Includes bibliographical references and index.
ISBN 978-0-393-06662-3 (hardcover)
1. AIDS (Disease)—Prevention. 2. Epidemiologists Biography. I. Title.
[DNLM: 1. Acquired Immunodeficiency Syndrome—prevention & control.
2. Acquired Immunodeficiency Syndrome—economics. 3. HIV Infections.
4. Health Policy. 5. Sexual Behavior. WC 503.6 P674w 2008]
RA643.8.P57 2008
614.5'99392—dc22
2007051396

ISBN 978-0-393-33765-5 pbk.

W. W. Norton & Company, Inc.
500 Fifth Avenue, New York, N.Y. 10110
www.wwnorton.com

W. W. Norton & Company Ltd.
Castle House, 75/76 Wells Street, London W1T 3QT

1 2 3 4 5 6 7 8 9 0

For Roger and Julie

CONTENTS

Map viii
Author's Note xi
Acknowledgements xiii

Preface: The Accidental Epidemiologist 1
1 Cooking Up an Epidemic 13
2 Landscapes of Desire 43
3 The Honesty Box 84
4 The Naked Truth 124
5 Sacred Cows 161
6 Articles of Faith 188
7 HIV Shoots Up 227
8 Ants in the Sugar-Bowl 269
9 Full Circle 301

Notes 327
Bibliography 347
Index 365

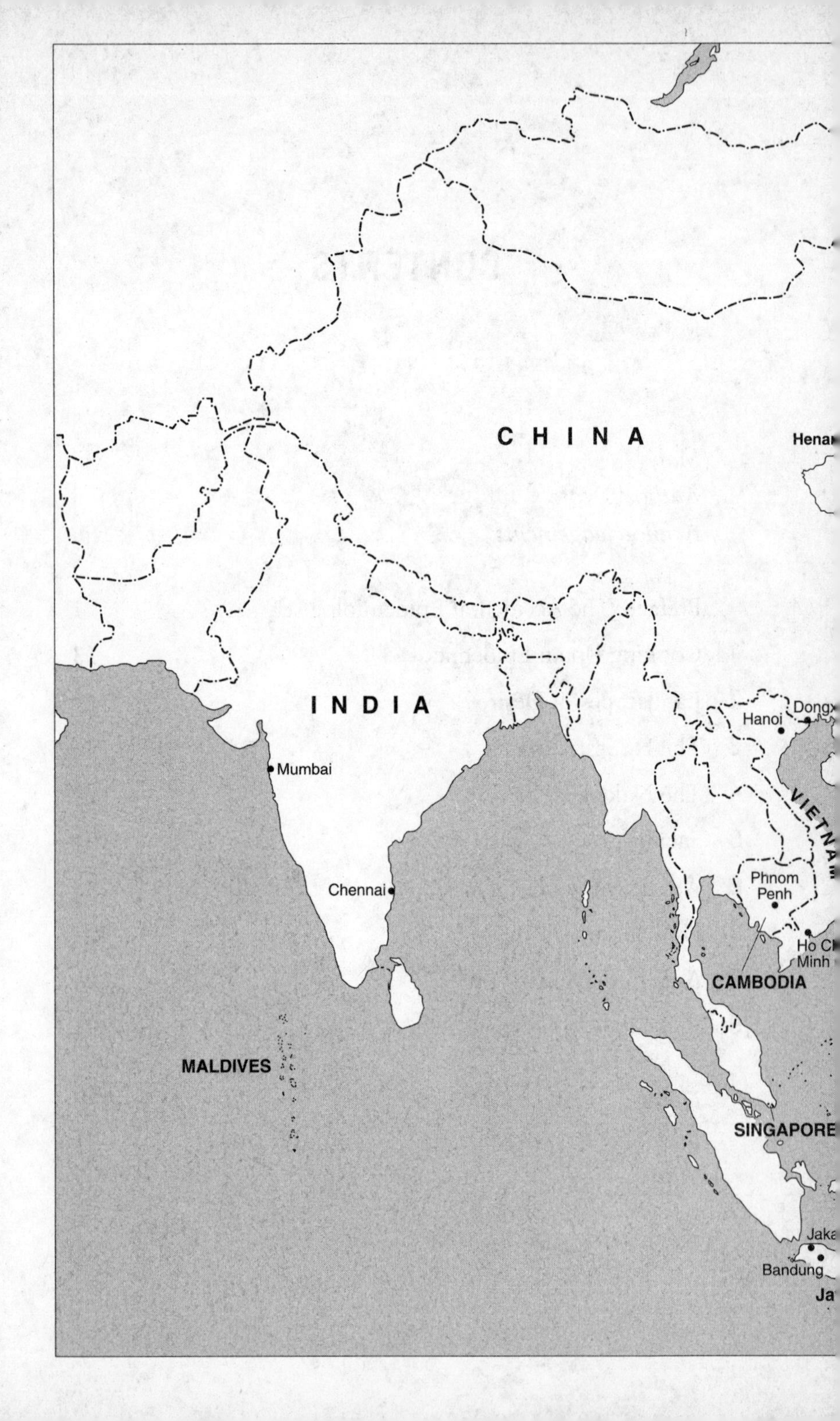
CHINA
INDIA
Mumbai
Chennai
Hanoi
Phnom Penh
CAMBODIA
MALDIVES
SINGAPORE
Bandung

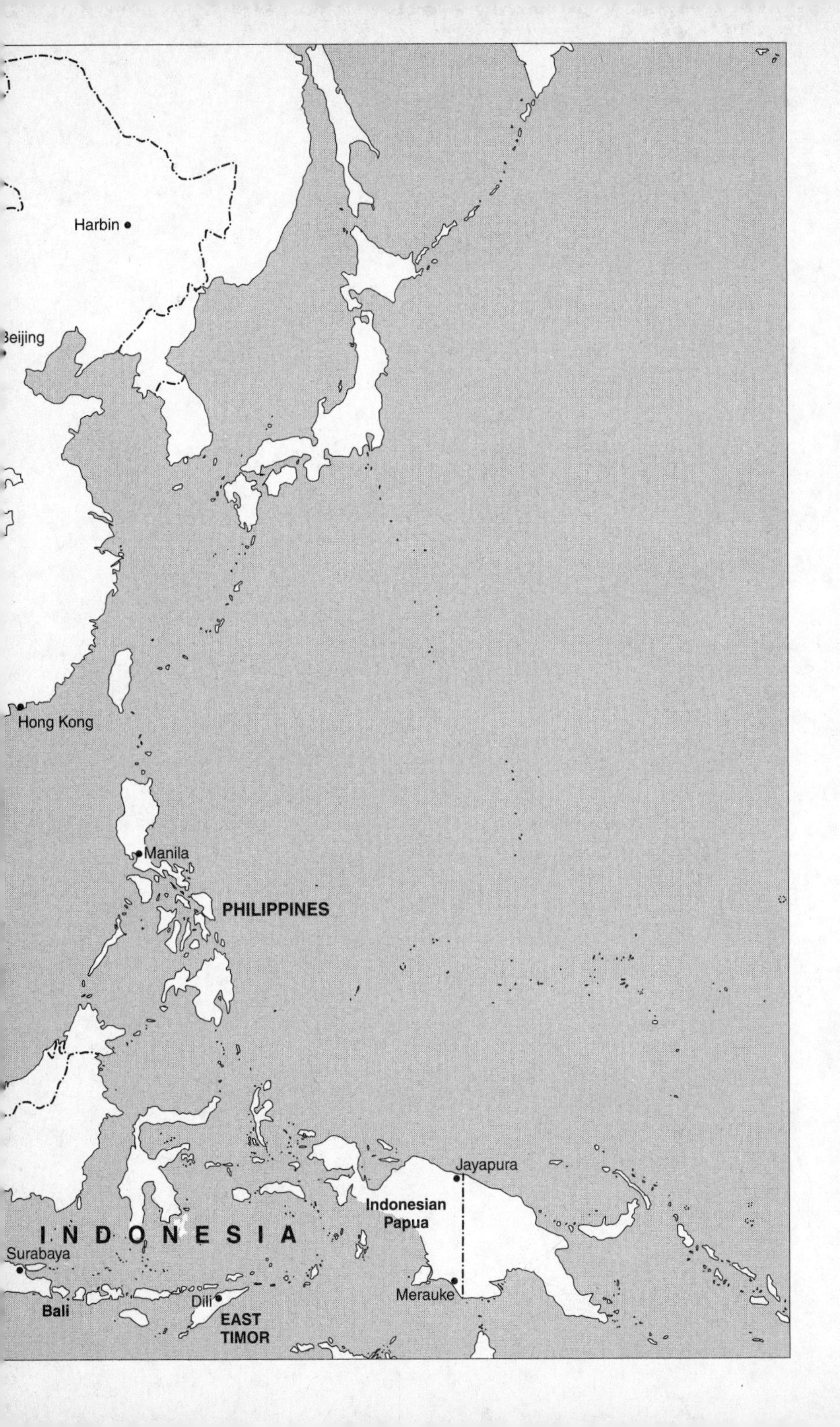

Harbin
Beijing
Hong Kong
Manila
PHILIPPINES
INDONESIA
Surabaya
Bali
Dili
EAST TIMOR
Jayapura
Indonesian Papua
Merauke

AUTHOR'S NOTE

This is a non-fiction book. The events and facts in it are recorded as faithfully as my memory and written and electronic records allow. Most people appear under their real names, or the names by which they introduced themselves to me. Sometimes this is a first name only (indeed many people in Indonesia only use one name).

Except where someone was speaking or writing on the public record, I have tried to seek people's permission to recount conversations we have held, or to reproduce their e-mails or written comments. In the few cases where I have not been able to trace people, I have generally changed their names.

Most of the documents referred to in the book, including the scientific references, can be accessed using the links provided on the book's website:

http://www.wisdomofwhores.com/references

The website also has illustrations, photos and my blog on topics that may be of interest to readers of this book. Comments are welcomed.

Elizabeth Pisani

London, November 2007

ACKNOWLEDGEMENTS

The Wisdom of Whores is the product of over a decade of debate, passionate disagreement, moments of enlightenment, misery and laughter shared with people in many countries and from many walks of life. Some I am not able to acknowledge by name, and some whose names I do mention will be uncomfortable with many of the ideas expressed in the book. I thank them nonetheless for contributing to my understanding and convictions.

My greatest debt of gratitude is to friends and colleagues in Indonesia – people who led me through brothels and bureaucracies with patience and good humour. They include Arizal Ahnaf, Ines Angela, Vidia Darmawi, Desi, Eris, Mohammad Noor Farid, Frankie, Molly Gingerich, Happy Hardjo, Naning Iswandono, Saiful Jazan, Balgis Koesnantah, Ratna Kurniawati, Lina, Ling Ling, Nana, Nancy, Palani Narayanan, Kharisma Nugroho, Brad Otto, Joy Pollock, Sigit Prihutomo, Chris Purdy, Haikin Rachmat, Resty, Made Setiawan, Fonny Silfanus, Arwati Soepanto, Stephen, Lenny Sugiharto, Aang Sutrisna, Bhimanto Suwastoyo, Wayne Wiebel and Dewa Wirawan. I learned much, too, from other colleagues at the Indonesian Ministry of Health's sub-Directorate of HIV/AIDS and at Statistics Indonesia, as well as at Family Health International's Indonesian programme, ASA. Thanks are especially due to Steve Wignall who led ASA with dedication, compassion and a healthy dose of ribald realism.

Besides making me laugh, Philippe Girault taught me a lot about the vagaries of fieldwork; Elizabeth Donegan did too. The unflappable Hari Purnomo was my guru for conflict resolution, and Penny Miller provided a shining example of the importance of persistence and attention to detail in turning good ideas into good programmes.

In Beijing, I was lucky to work with the epidemiology division of the National Centre for HIV/AIDS/STD Control. Thanks to Lu Fan and his staff, especially Chaganhuar and Wang Liyan who were always willing to burn the midnight oil in the name of research. I'm also grateful to Cheng Feng, He Jinglin, Bert Hoffman, Liu Qian, Meiling, Joel Rehnstrom, Wang Lixue, Wu Zunyou and Zhong Li.

Some say it is more important to choose a good boss than to choose a good job; I have been lucky to have both. Jeanine Bardon ran FHI in Asia with an unbeatable mixture of good humour and steely determination to build programmes that actually improve lives. She ran an organization stuffed with wonderful people who taught me many things. Thanks to all of them, especially Sara Hersey, Nancy Jamieson, Nigoon Jitthai, Mike Merrigan, Graham Neilsen, Dimitri Prybylski, Boyet Roquero and Ganrawi Winnithana. Steve Mills and Tobi Saidel were my first friends at FHI; they taught me about behavioural surveillance and pulled me into the fold, and for that as well as good friendship I thank them.

I've had many good bosses in my wanderings as a consultant: Ties Boerma, Txema Calleja, Michel Carael, Peter Ghys, Rob Moodie and Bernhard Schwartlander have all indulged my argumentative nature and defended my plain prose against UN-creep. Other friends and colleagues in the AIDS mafia include Andrew Ball, Chris Castle, Kevin De Cock, Don Des Jarlais, Isabelle de Vincenzi, Geoff Garnett, Robin Gorna, Simon Gregson, Daniel Halperin, Noerine Kaleeba, John Novak, Wiwat Peeratapanapokin, Cyril Pervilhac, Deborah Rugg, Swarup Sarkar, Karen Stanecki, John Stover, Jim Todd, David Wilson, Ray Yip and Basia Zaba. I have argued stubbornly with most of these people over the years, and with some I still disagree on many points, but I have always come off the wiser from our encounters. I'd especially like to thank Tim Brown, who shares my brain, and Neff Walker, who simply makes the AIDS field a more fun place to work.

It goes without saying that I thank everyone who appears in the book for letting me tell their tales. First among equals is David Fox, who has been in equal parts generous and courageous in allowing me to chronicle an unhappy episode in our largely happy history.

Several people were kind enough to read and provide detailed comments on drafts of all or parts of the book. Thanks to Chris Archibald, Peter Baldwin, Andrew Ball, Jeanine Bardon, Julie Bates, Claire Berlinski, Karen Broughton, Sophie Campbell, Michel Carael, Adrian Cox, Tim France, David Gordon, Robin Gorna, John McGlynn, Jack Molyneaux, Palani Narayanan, Mead Over, Sophie Pinwill, Gray Sattler, Nathalie Steiwer, Jonathan Swire, Robert Templer and Jack Tenison. Sinead Worrall brought a whole new perspective to the book. Keith Hansen spent untold hours pointing out logical fallacies, highlighting clichés and suggesting improvements in content and structure. He managed to do all of that while still being a fount of energizing encouragement. Jimmy Whitworth has entrenched his position as the epidemiologist I most admire by taking the time to read and comment thoughtfully and amusingly on every chapter of this book. For that and for a decade of inspiration, many thanks. Chris Green gets my vote for 'world's best activist'. Through careful readings and forthright comments Chris has tried to keep me both honest and human. I apologize if I continue to fail on either front. I can only say that the book has been vastly improved by the contribution of all of these readers.

I've been as lucky as it is possible to be as an author. The rather improbable idea of trying to sell an engaging book about a fatal disease was enthusiastically swept up by Felicity Bryan. I am indebted to Daphne Park for introducing us, and to Felicity, George Lucas, Andrew Nurnberg and their teams for match-making me with such wonderful publishers. Sara Holloway at Granta and Alane Mason at W. W. Norton prodded me kindly but pointedly in directions that made the book better; they have been a pleasure to work with. David Graham, Brigid MacLeod, Pru Rowlandson and Alex Cuadros have also provided good ideas and encouragement.

The Wisdom of Whores exists because Olivia Judson willed it into being. My debt to her is incalculable. Nicola Bullard provided a thesaurus while I was writing the book; it was part of a package that included a home in Bangkok, editorial advice, champagne and friendship; I value all of these, but especially the last. Havens of encouragement and good cheer were also provided at critical moments by Jack Tenison and Susannah Fiennes, by Jonathan Stull and by Lilo Acebal, whose burning enthusiasm for life has infected me for two decades. Claire Bolderson and Andrew Wilson have allowed me to be a perpetual squatter in their lives and homes since the beginning of time. They have housed, fed, entertained, provided secretarial support, edited, cajoled and uplifted. It seems unlikely that I will ever be able to thank them adequately. The Damais family in Jakarta have provided similar sustenance in my adopted home. *Terima kasih*.

Finally, I'd like to thank my own family. My brother Mark has held my hand from the start of this project, commenting on proposals and drafts with the brutal but indispensable honesty that only a sibling can muster. Annoyingly, he is always right. Mark has been editor, fact-checker and IT adviser to a sister who works on short deadlines in a different time zone; he doesn't get much sleep. He has been a believer in this book even when I was agnostic-tending-to-atheist. And he's made it all a lot of fun. I can't thank him enough.

My parents met when my father was hitchhiking around the world and my mother was hitching around Europe. Perhaps not surprising, then, that they bred into me curiosity, a love of travel and a tendency to talk to strangers in immigration queues. The result is that they have a largely absent daughter with a confusing tendency to change countries and occupations. Yet they have been steadfast in their support for all of my ventures – from that support I draw strength and happiness. Thank you both.

THE WISDOM OF WHORES

PREFACE

The Accidental Epidemiologist

When people ask me what I do for a living, I say, 'Sex and drugs.' I used to say I was an epidemiologist, which is also true. But most people looked blank. Epi— what? Perhaps something vaguely distasteful to do with skin.

Saying I do sex and drugs saves me explaining that epidemiology is the study of how diseases spread in a population. It saves me from the social suicide of admitting that I am a scientist, a glorified statistician, a card-carrying nerd. And it is a good conversation starter. Everybody has something to say about sex and drugs.

I've discovered that fact in a decade of researching sex and drug injection around the world. Not an obvious career choice for a nice Catholic girl, perhaps. In fact, not a career I even knew existed for most of my life. I became an epidemiologist by accident.

As a child, I followed my corporate exec parents around Europe, learning cuisines and languages, wandering through flea markets and billiards halls. When I was fifteen I went to visit a school friend in Hong Kong. We threaded our way through alleyways slithering with eels in great plastic tubs, we dodged bow-legged hawkers shuffling between swinging baskets of lychees, we stuck our tongues out at lollipops of dried terrapins, crucified on their bamboo sticks. We gyrated with rich kids in flashy nightclubs, and peeked into tawdry girlie bars. Then we grew bolder, walking into the girlie bars, ordering beers and chatting to bored sailors, bored bankers, bored hookers – anyone who would chat. I discovered that everyone has something interesting to say. I was hooked on Asia, hooked on nightclubs and girlie bars, hooked on chatting to anyone who would chat.

I took a degree in Chinese and set off back to Hong Kong, bouncing into a job as a foreign correspondent with Reuters news agency – about as fab as it gets for someone who just likes to chat. I reported on liberation wars and stock market booms and the massacre of hopeful students by hopeless despots. I visited brothels and orang-utan sanctuaries and military graveyards. I went to mass with the pope and school with rice farmers in the paddy fields. I learned some new languages and chatted with thousands of interesting people, and I got paid for it. I loved it, but I did grow tired of trying to reduce human experience to 600 words on a two-hour deadline.

Elbowing through the crowds of Hong Kong, New Delhi, Beijing, Jakarta, I became interested in the politics of population control. Sex and birth, health and death, priests and condoms, forced vasectomies and contraceptive 'safari camps' (line 'em up on fold-out camp beds and stick in the coils). The different approaches taken by the mega-nations of Asia would determine their own future, and perhaps the world's. There were days when that seemed more interesting than writing reports on the dollar–rupiah forex market. On top of that, I'd fallen for a boy who had moved to London and had a giant, generously rent-free house. So in 1993 I quit the job, moved to London and signed up for a Masters degree in medical demography. I wasn't quite sure what it entailed, but I knew I'd learn some number-crunching and have luxurious hours in wood-panelled libraries just thinking. Deadlines be damned.

On the wall outside the wood-panelled library of the London School of Hygiene and Tropical Medicine danced giant gilded sculptures of mosquitoes, fleas and tsetse flies – the vectors of the diseases the School specialised in in the heyday of colonialism. The library itself seemed a relic from those days, its leather armchairs and leather-bound volumes musty in the dappled mid-afternoon sunlight. Outside the oasis of the library swarmed doctors and virologists, lab technicians and statisticians from every part of the globe. I was moderately numerate – all those stock market and forex stories – but my exposure to other sciences ground to a halt at fourteen, when I took my last biology exam. A degree in classical Chinese and five years in graveyards and paddy fields did not seem like adequate preparation for a new life as a scientist, I thought, as I picked up the folders for the two courses required of every student at the school. Statistics and epidemiology. Epi— what?

In the first lecture, we 'reviewed' all the major study types. For example, in the case-control study you find a group of people with a disease, and then look for people who are much the same but without the disease. You compare the two groups to see if they have different risks. It's a relatively cheap method, but it doesn't tell you much about the order in which things happen. I can't remember all the examples used in the lecture, but let's say you want to look at causes of depression in women. You start with 600 depressed women, find another 600 who match them in age, ethnicity and educational status, and then ask them all about their lives. Let's say you find out that women who are depressed are six times more likely not to have had sex in the last year as women who are cheerful. That means if you're not having sex you get depressed, right? But hang on, couldn't it be that women who are moping around looking miserable don't get laid much?

Perhaps you'd be better off with a cohort study. You start off with several thousand women who are perfectly happy. Then you follow them over time, recording their behaviours, and see which of them get depressed. If you find that women who have sex are less likely to become depressed than women who aren't getting any, it suggests it is the lack of sex that causes the depression, not the depression which stops you getting laid. You can throw out the 'misery guts' theory and recommend more good sex as an intervention to promote mental health.

This may have been a review to most people there, but it was all new to me. The barrage of 'facts' that we see in the newspapers every day took on a new perspective. Red wine is good for you. No, no, red wine is bad for you. Well, actually, red wine is good for you but only if you are white, over sixty, and drink less than 65 millilitres a day. Even for scientists, the stats are not straightforward. Suddenly, epidemiology began to look interesting.

At the end of that first lecture, the professor asked a question. Why was there a fourteen-year gap between the first case-control study showing a strong association between smoking and lung cancer, and the first US Surgeon General's report on the dangers of smoking?

Stony silence from the highly educated doctors and technicians in the room, men and women who were adding a public health qualification to an existing wealth of medical experience. Maybe this was because it was the first lecture of the year and people were shy. I was not a doctor. I did not have an existing wealth of medical experience. I had not had any scientific education in twenty-five years. But I was not shy. A journalist's work depends on a willingness to ask questions of people who are better informed and more powerful than you. It depends on regarding nothing as sacred and everything as open to question. I was by far the least qualified of the 300 or so people in that echoing lecture theatre, but I was full of been-there-done-that bravado. I stuck up my hand.

'You're asking the wrong question,' I said.

Even I was aware that the air in the lecture theatre had suddenly turned heavy. Heavy enough to crush the bravado. I blundered on, more doubtful now.

'Surely, the key question is: how much money did British American Tobacco and Philip Morris give to US Senate campaigns in that fourteen-year interval?'

Immediately, there was a shower of laughter and the air cleared. A forest of hands shot up, everyone competing to explain in technical terms that I only partly understood: case-control studies are subject to recall bias, case-control is not the most appropriate method for looking at causes of death, what is really needed to confirm the findings is a cohort study that follows both smokers and non-smokers over time, and and and . . .

All of these answers were correct, of course. But did that mean the Big Tobacco answer was wrong?

Science does not exist in a vacuum. It exists in a world of money and votes, a world of media enquiry and lobbyists, of pharmaceutical manufacturing and environmental activism and religions and political ideologies and all the other complexities of human life.

There's plenty of evidence that a lack of sound science was not the only thing that dragged down action to discourage smoking. The 1950 study showed that there were twenty-one times more non-smokers among men who did not have lung cancer than among men who did have lung cancer. If that's hard to follow, it's because one of the downsides of a case-control study is that you can't say A leads to B. So yes, you don't want to make public health policy just on the results of that one study. But another 7,000 studies showing similar and stronger results were published before the US Surgeon General risked the wrath of the rich and powerful tobacco companies by saying that smoking is bad for you. Well, he didn't actually say bad for you. He said 'a health hazard of sufficient importance to warrant appropriate remedial action'. Nowhere in the Surgeon General's 387-page report did he venture what 'appropriate remedial action' might be.[1]

The report was released during a carefully orchestrated 'lock-in' of accredited journalists, held on a Saturday morning. 'The date chosen was a Saturday morning to guard against a precipitous reaction on Wall Street,' according to an official history of the 1964 Surgeon General's report posted on the US Centers for Disease Control (CDC) website. Which sounds to me like an admission that science, case-control study or no, is not the only thing politicians consider when making decisions about public health.*

The more I thought about it, the more I liked epidemiology. It's actually not unlike investigative journalism. You need to ask the right questions of the right people. You need to record the answers carefully, analyse them correctly and interpret them sensibly, and in context. And you have to communicate the

* This statement, along with a number of others that might be deemed unfriendly to Big Tobacco, was removed from the CDC website in early 2007. To view both the original web page and its replacement versions, and to find links to most of the documents, papers and illustrations mentioned in this book, see http://www.wisdomofwhores.com/references. For a rigorous account of tobacco industry lobbying practices, see Glantz et al., 1996, available online at http://ark.cdlib.org/ark:/13030/ft8489p25j.

results clearly to people who might do something about them. Journalism (day-to-day news journalism, at any rate) is frustrating because you don't always have the time or the tools for thorough analysis. Epidemiology gives you that. But it seemed to me that epidemiology often falls at the last hurdle: the communication.

I soon learned that the world of epidemiologists, perhaps like any professional *demi-monde*, is deeply divided. On the one side are those who believe an epidemiologist's job is to do good science. End of story. Turning good science into sensible policy is someone else's job. This camp, which apparently includes the editors of the scientific journal *Epidemiology*, actually believes that it is wicked for epi-nerds to get involved with policy, because it compromises their scientific neutrality.[2]

The other camp believes that epidemiology and public health are inseparable. Public health is not glamorous, and it is not especially well paid. You work in public health because you want to save a lot of lives. If you're going to do that effectively, you can't stop at the perfect study design, or even at the publication of your perfect paper in *The Lancet* or the *New England Journal of Medicine* (the dream of epidemiologists in both camps). An epidemiologist is a scientist, yes, but in the public health camp that's not enough. Something that works in the lab but doesn't work at the ballot box might be good science, but it is unlikely to get translated into good public health. So you have to *do* good science, and then *sell* good science.

An idea that kept gnawing at me as I lounged in the library's leather armchairs: we could save more lives with good science if we spent less time worrying about publishing the perfect paper and more time lobbying, more time schmoozing the press, more time speaking in the language that voters and politicians understand. If we behaved more like Big Tobacco, in fact.

I could have chosen to work on malaria, or dengue fever or maternal mortality. But if your real interest is the shadowy area where science does battle with politics, you want to go for the issue that makes politicians most squeamish. And in the mid-1990s, it seemed that issue was AIDS.

AIDS had first blundered into my consciousness in New York in 1981, when I was working in a fashion advertising agency for a year before going to university. Headlines in the *Village Voice* newspaper screamed about GRIDS – Gay Related Immune Deficiency Syndrome. At first, the screaming fell on deaf ears. At the weekends, when I'd sometimes follow the city's gay swarms out to the beach at Fire Island, I'd have to tread carefully to avoid tripping over men entwined with one another in the dunes. Entwined couples were much more common than condoms on Fire Island in those days. Within a year, GRIDS had worked its way into New York's consciousness. Cafés grew hushed when yet another skeletal figure shuffled in. Drinks with friends in the city's gay bars were often interrupted by volunteers from the Gay Men's Health Crisis, handing out leaflets and condoms. By that time drug injectors, haemophiliacs and Haitians had been added to the list of 'victims', and the disease was renamed Acquired Immunodeficiency Syndrome, or AIDS. The very word AIDS seemed to strike terror into the hearts of politicians. Ronald Reagan presided over the emergence of the epidemic in the United States. He witnessed American and French scientists race to identify the virus that caused it. He saw his old acting buddy Rock Hudson waste away. But it wasn't until September 1985 that he managed to say the word AIDS in public.[3] Rock Hudson died of AIDS two weeks later.

The same year, a wasting disease ravaging Uganda which the locals called 'Slim', was identified as AIDS, and every region of the world reported at least one case. Britain had clocked up 275 cases by the end of 1985. The tabloids were hysterical about 'Acquired Immoral Deficiency Syndrome', and public bodies such as the Blood Transfusion Service were tiptoeing towards the truth with genteel warnings about the dangers of 'intimate contact'. When reported cases topped 1,000 in 1987, the government grew less squeamish. Through every letter-box in the country, all 23 million of them, fluttered a leaflet giving chapter and verse about HIV, condoms and safe sex.[4]

AIDS didn't make it to the London School of Hygiene's curriculum until I was there, in 1994. By then we knew that almost all HIV-infected adults got their infection when having anal or vaginal sex, or while injecting drugs with shared needles. Infected blood products could spread the disease, though that was on the wane. And mothers could pass HIV on to their infants, in the womb, at birth or while breastfeeding. We knew that in rich countries, AIDS was a disease of gays and junkies, of prostitutes and their clients. Those groups were affected in some poorer countries, too. But in black Africa and the Caribbean, HIV didn't seem so picky. It seemed happy to target just about anyone who had sex.

Sex, drugs and plenty of squeamish politicians. AIDS was the disease for me.

That choice shaped the next ten years of my life. It set me up for a decade of adventure, discovery, hilarity, hope, disappointment. It allowed me to explore the guts of worlds I had barely known existed. From prostitutes, rent boys, pimps and clients I learned about the sex trade. Addicts, cops and rehab workers taught me about the parallel universe of drugs. Perhaps the hardest world to find my way around was the AIDS industry itself, a world where byzantine international bureaucracies fight turf wars with one another, with pharmaceutical giants, with activist NGOs. A world where money eclipses truth.

AIDS was not a fashionable subject at the start of my career in public health, the starting point for this book. It was assuming pride of place as the number one killer of young adults in more and more countries, but many people still preferred to close their eyes to it. Our first task was to draw more attention to the disease, to persuade governments to do something to prevent their growing HIV epidemics, and to find cash to help them do it. I immersed myself in these tasks, in the company of a colourful band of characters crowded into the corridors of an upstart UN agency in Geneva. We painted glittering portraits of prevention success and thundered about the tragic consequences of failure. We manhandled estimated infections and manipulated maps. We did well at drawing attention and finding cash, but appallingly badly at persuading governments to do the right thing.

Perhaps if we had better information? I threw myself into the task of helping countries understand their epidemics better. We wrote guidelines and toolkits, manuals and handbooks, instructing people how to measure their epidemics better. Then I took the guidelines off to Asia and road-tested them in Indonesia, in China, in East Timor, in the Philippines. I encountered a world of women with penises who sell anal sex to men who are completely heterosexual. I found men who buy sex from women and sell it to men. I found heroin addicts who fly aeroplanes and Muslim fundamentalists who run protection rackets for brothels.

I learned a lot about the warts that exist on the underside of all those health statistics you see in your newspapers every day. They make things seem so simple, those numbers, but they're boiled up out of cauldrons of uncertainty, of best guesses, of spilled samples, of errors corrected on the fly. When we did begin to produce HIV statistics that we thought were fairly solid,

I found that they often didn't support the conventional wisdoms of the AIDS world. Was it true that the HIV epidemic in Asia would soon explode like that in Africa? No. Were most prostitutes the victims of trafficking or coercion? No. Would more premarital sex put more young people at risk for HIV? Quite the reverse.

Bit by bit, we got a better idea of what was really going on, what put people at risk of HIV in Asia, and what we should be doing about it. We imitated Big Tobacco as best we could, packaging up the data for different politicians, lobbying various interest groups, massaging the media. And still, we found it really hard to get governments to do the right thing.

I started to take a look around at the wider picture. And everywhere, I saw the same thing. We were collecting more and more really good information, and then not acting on it. Two things were getting in the way – ideology and money. In the AIDS industry, we have too much of both.

Most ideologies are religious or political, but we also stumble over the politically correct convictions of the AIDS activists who led the initial charge against HIV. Whatever their source, these ideologies influence what we do about sex and drugs and determine how we do it. Money, of course, follows the dominant ideologies. But it also sucks in people who don't really care about the problem, who are truly queasy about sex and drugs, but who want some of the cash. When AIDS was unfashionable, we had thought that more money would make it easier to do the right things to prevent a wider epidemic. Now that it is a boom industry, it has become clear that money can actually be an obstacle to doing the right thing.

This book tells the tale of the worlds I have encountered during the boom years of the HIV industry. It is rooted firmly in my experience, and the experiences of the people who have held my hand and walked me through the brothels, shooting galleries and boardrooms where the future of AIDS is shaped. Unlike many books about AIDS it does not focus primarily on Africa, where two-thirds of HIV infections have been transmitted so far. Most of the characters, the stories, the data come from Asia (and particularly Indonesia) where I have done most of my research. You will not find apocalyptic accounts of an AIDS tsunami about to engulf the world's most populous continent. But you will find that if you count up just the female, male and transgender sex workers of Asia and their clients, and you add in Asia's drug injectors and the men who cruise for new partners in the continent's blossoming gay scene, you'll come up with a number that is not all that far off the entire adult population of sub-Saharan Africa. And you'll find, too, that the HIV epidemic in Asia looks very much like the epidemic in the West, in Latin America and in Eastern Europe. The people I'll introduce you to may live in exotic subcultures, but their wisdoms, the lessons we learn from them, are often universal.

Perhaps if I had known in the mid-1990s what a roller-coaster of triumph and despair HIV would turn out to be, I'd have chosen to work on maternal mortality. Perhaps if I had known how people at the next dinner table would react when my table traded stories about drinking foreskin soup, I'd have gone for dengue fever. Or if I had known how often I would be taken aside by a friend or acquaintance in meltdown because they had just tested HIV positive (or their brother had, or their girlfriend, or their boss). Perhaps if I could have foreseen how our lobbying successes would lead to billions of dollars of taxpayers' money being shovelled down an ideological drain, I would have chosen differently.

But in the mid-1990s I didn't know any of this. A career in sex and drugs? It seemed like a good idea at the time.

1

Cooking Up an Epidemic

How do you launch a career in sex and drugs? Like any other career: try to be in the right place at the right time.

In 1996 the right place was UNAIDS, a brand new and slightly amorphous 'joint programme' that was supposed to stop the United Nations agencies bickering about whose job it was to deal with AIDS.

A decade earlier, the World Health Organization (WHO) had tried to stamp its authority on the epidemic by establishing the Global Programme on AIDS. It made the fatal mistake of trumpeting the fact that Human Immunodeficiency Virus (HIV) had serious consequences for social and economic development. 'Development? Then it's our business!' said the United Nations Development Programme. 'Economics? That's us!' said the World Bank. UNICEF fretted that children were involved, the United Nations Population Fund ruled over contraception and thus condoms. The United Nations Educational, Scientific and Cultural Organization found that AIDS ticked all the boxes in UNESCO's own title. None of these agencies was actually doing much about AIDS, but they didn't want those medics at WHO muscling in on their turf. WHO spent so much time defending itself from the slings and arrows of other agencies that its own work crumbled.[1]

In the end, the countries that fund the UN got fed up and decreed that the Global Programme on AIDS should be dissolved. WHO must get together with the other five agencies who then staked a claim to AIDS and form a joint programme, UNAIDS, which was supposed to coordinate the efforts of the different agencies and reduce duplication. It was to make sure that when they serenaded governments about AIDS, the different members of the UN family sang the same tune. It was also supposed to act as a clearing house for statistics, providing a single set of 'official' estimates about HIV infection and AIDS. In 1995 Belgian virologist Peter Piot was given the task of putting the programme together.

Piot was like a nursery school teacher trying to get all the children to play nicely together in the sand pit. It wasn't easy, because WHO, the biggest kid in the AIDS class, was still sulky about having its toys taken away and given to the other agencies to play with. But Piot was clever. He ignored most of the time-serving bureaucrats who sniffed around the new programme looking for nice, secure jobs. Instead he sought out people who did real work in the real world – UN virgins, mostly. Like Piot himself, they were people who had come from laboratories in Congo, from VD clinics in Zimbabwe, from condom promotion programmes in Thailand, from outreach work with transvestite prostitutes in the Bois de Boulogne. They signed up because they knew what works in the messy landscapes in which HIV spreads, and they wanted to share that knowledge with other people around the world. They really wanted to prevent what we knew was an eminently preventable disease, to make sure that the people we had already failed got some care. These mavericks plunged into a world of incomprehensible acronyms, of inter-agency task forces, of meetings whose main purpose was to call another meeting. Eventually, the UNocracy would wear many of them down; they would either get out or sell out. But in those start-up years UNAIDS was a place of drive and optimism.

It was into this world I drifted, keen to contribute my shiny new public health skills to the cause. I didn't actually want to live in Geneva, though. The boy I had followed to London was by then in another giant, generously rent-free house, this one in the Kenyan capital of Nairobi. When he wasn't off doing his war correspondent thing at one of the region's many conflicts, we were happily ensconced together. Nairobi to Geneva seemed like a long commute, but it turned out that it was really quite normal by the standards of the UN. A job with the UN is essentially a job for life, so agencies are careful about hiring new full-time staff. This means a huge proportion of the work is done by 'short-term consultants' – people who are happy to sign on and pitch up in Geneva for a week or a month or a year at a time. People think of the UN as a giant gravy train, but the train conductors have sophisticated ways of keeping the working passengers out of the carriages that confer benefits and pensions. I know people who have been 'short-term consultants' for eleven years or more. I examined the skills I had to sell. I could do nerdy things like run a regression analysis, use computer models to predict infectious disease, interpret the results of sexual behaviour surveys, code data, even fake cost–benefit calculations. I sold all of these skills at one time or another, mostly to UNAIDS in Geneva but also to WHO, the World Bank, the US government, the British government – whoever would pay, really, as long as I could go back to my boy in Nairobi at the end of the day, or the week, or the month.

To my surprise, the skill that was most in demand was not one

I'd battled to learn in graduate school but one I'd absorbed in the newsrooms of London, Hong Kong and New Delhi. I could write plain English. Fast, to order, about more or less anything.

Fast was important. There was an enormous sense of urgency at UNAIDS in its early years. Everything seemed to happen in a panic at the last minute. The sober WHO lifers with whom we shared our corridors must have been startled by UNAIDS staff with nose rings and Jean-Paul Gaultier shirts, running around and yelling, 'I'm coming, I'm coming. But I can't find the anal sex stuff.'* We sat in our offices and talked passionately about hookers, queens and junkies. In those days, when there was still more passion than politics and where many people in the room fell into at least one of the groups we were talking about, we bulldozed happily through the minefield of language.

In polite company we weren't allowed to say queens, obviously. In fact, we couldn't even say gay, because a lot of men who have anal sex with one another don't think of themselves as gay, and we care about the behaviour, not the identity. So it became men who have sex with men, or MSM for short. Oh, but wait, some of the people with XY chromosomes who have sex with other people with XY chromosomes don't think of themselves as men. Luckily, we didn't have to change acronyms. It

* The culture clash became even more severe when part of UNAIDS moved down the road to share a building with the good folks of the World Council of Churches.

† I leave those who think I am exaggerating to make sense of this footnote from a recent UNAIDS document entitled 'Men Who have Sex With Men': 'While we use the term "men who have sex with men" here it is within the context of understanding that the word "man"/"men" is socially constructed. Nor does its use imply that it is an identity term referring to an identifiable community that can be segregated and so labelled. Within the framework of male-to-male sex, there are a range of masculinities, along with diverse sexual and gender identities, communities, networks, and collectives, as well as just behaviours without any sense of affiliation to an identity or community.'

became Males who have Sex with Males.† IDUs switched identity too, from intravenous drug users to injecting drug users, because some people inject into muscles, not veins. 'Prostitute' was considered a judgement rather than a noun, except among French speakers, so that was out. Selling sex is a legitimate profession, and we can recognize it as such by calling prostitutes commercial sex workers (CSWs). But wait, a lot of women sell sex not for profit or even for cash, but for survival, for school fees, to pay for Dad's visits to the doctor or the beer hall. So it is not necessarily commercial. Just 'sex workers' then. But SW doesn't trip as easily off the tongue as a three-letter acronym. So let's make it FSW, female sex workers. Later, Washington decided that it didn't recognize sex work as a legitimate profession after all, and tried to take us back to prostitute.

Eventually it all got so complicated that someone decided to pour the alphabet soup into a new bowl, coming up with acronyms that avoid us even having to think about male–male sex or drug injection. Current favourite is MARPs, 'Most At Risk Populations', but I gather the PEHRBs, 'Persons Engaged in High Risk Behaviours', are waiting in the wings.[2] As if my colleague in the Jean-Paul Gaultier shirt would rather be called a PEHRB than gay.

It gets sillier. I once received an e-mail from my colleague Graham Neilsen about some technical guidelines I had written under the engaging title: 'Estimating the size of sub-populations at risk for HIV: Issues and methods'. Poor Graham had been press-ganged into the language police and had to plod through this dry tome. 'In the committee-like process of reviewing the cover of your book, the question of the potentially negative connotations of the term "sub-populations" was raised. It is a tad subtle but the term may cause distress to some stigmatized groups by suggesting that they are somehow "beneath" mainstream society etc.'[3] This from a man who was described on Australian national television not as a doctor specializing in the treatment of sexually transmitted infections (STIs, by the way), but as a 'Dildo Expert'.

If you pay taxes to any member state of the United Nations, you have paid for people on five-figure salaries to sit around in five-star hotels for days discussing these important issues.[4] But in the privacy of their offices, the same people often use shorthand that many would consider offensive – prostitute, hooker, john, queen, fag, whore, rent boy, trannie, junkie, smackhead – the terms that real people use to refer to themselves in daily life. In the context there is absolutely no offence intended. Certainly, no offence is intended when I use these words in this book.*

Beating it up

The urgency felt by the mavericks that Peter Piot had pulled in to UNAIDS was not, initially, shared by the UN agencies that the programme was supposed to be representing. The United Nations tends to mirror the obsessions of its largest funders and voters. Since the panicky days of the mid-1980s, when even Ronald Reagan was forced to look AIDS in the face, the epidemic had slipped down the agenda of industrial countries. By the mid-1990s, it had become clear that HIV was not going to charge through the 'general population'. Gay communities had effective prevention programmes in place, and the countries that cared about the health of drug injectors, such as Australia,

* I am aware that many people feel that only those who are members of a particular group have the right to appropriate its slang, and I apologize to them for my intrusive use of language.

Canada, Switzerland and the UK, were keeping the lid on injecting epidemics too. We'd seen HIV storm through the sex trade in Thailand in the first few years of the 1990s, but the government there had acted quickly and comprehensively to bring things under control. People were vaguely aware that things weren't going quite so well in Africa, but in the country where we had the best data, Uganda, there were encouraging signs that a corner was being turned. Overall, the feeling was one of inertia. Rich countries were no longer panicking about AIDS at home, and the few poor countries that were collecting good information seemed to be coping. Most countries didn't want to hear about AIDS at all, let alone think about the sex and drugs that spread it.

Those of us who tracked the epidemic more closely thought this complacency misguided. HIV infection had reached achingly high rates among drug injectors in countries as diverse as the United States, Burma, Nepal and Ukraine. In parts of Thailand and India the virus had shown that it could rage through the sex industry, infecting up to half of prostitutes within a couple of years. In Botswana, Malawi, Uganda, Zimbabwe and Zambia more than one adult in ten was infected with HIV by 1995, and AIDS was the top killer of young adults in dozens of other countries. At the time, we knew that there was plenty of risky sex going on throughout sub-Saharan Africa, as well as in Brazil and several of the larger Caribbean islands. We also knew that drug injection was on the rise in some parts of Eastern Europe, and that male–male sex was pumping HIV into several niches in Latin America. We knew that Asia's sex industry was huge – much bigger than its condom industry. There were far more signs of risk than there were signs that anyone was doing anything about it.

We'd seen HIV blaze down the right-hand side of Africa from

Uganda to South Africa, and it seemed quite plausible that it would blaze back up the left. Now we could see sparks flying into petrol-tanks of risk everywhere else in the world. Populations of drug injectors, of gay men, of sex workers and clients were beginning to go up in flames. But no one seemed to be getting out the fire hoses. In 1995 the world spent less than US$250 million trying to extinguish the HIV epidemic. These days, Americans spend over eight times that amount, two *billion* dollars a year, just on Botox injections to extinguish their wrinkles.[5]

We were never going to be able to extinguish HIV with US$250 million. Not practically, and not politically, either. In most poor countries, governments were already closing their eyes to the early warning signs of HIV. Some had kept their eyes shut for so long that AIDS was now an obvious problem, clogging up hospitals and graveyards. Those were the ones that were getting rewarded with a chunk of the paltry AIDS funding. Elsewhere, we were expecting often shaky governments to take politically unpopular measures against an as yet unseen disease, and we couldn't even bribe them to do it with the promise of funding or technical help.

The first thing we needed to do was get more money for HIV prevention. That meant convincing rich countries that they should worry about AIDS in poor countries. Once we'd done that, we could worry about cajoling the governments of poorer countries to spend the money sensibly, to damp down HIV and to cushion the blows that it deals to people's health and to their families. How would we convince rich countries to stump up cash for AIDS? By making them feel the heat of the epidemic as it blazed across one continent and flared up in others, by painting a compelling picture of the devastation left behind, by infecting them with the same urgency that coursed through the corridors of the upstart UNAIDS.

The first attempt to inspire people with the horrors that were in store unless we took bold action right now began thus:

> By 30 June 1996, 1,393,649 cumulative cases of AIDS in adults and children had been officially reported by countries – an increase of approximately 19% over the 1,169,811 reported by 30 June 1995.[6]

If you made it through five pages of arid statistics, you might discover that one in six pregnant women was infected with a preventable, fatal disease in some countries. UNAIDS itself was estimating that 28 million people around the world had already been infected. That year, there were enough new infections every day to fill twenty jumbo jets. And we thought we could inspire people to action with 'By 30 June 1996, 1,393,649 cumulative cases of AIDS . . .' I'm not sure who wrote that joint WHO/UNAIDS report, but I could see why UNAIDS thought it could use some writing skills.

The next report was published the following year for World AIDS Day, 1 December 1997, and it was largely in my hands. The numbers were more dramatic – estimates had leapt to 30.6 million by then – but the language was only a little less turgid:

> New estimates show that infection with the human immunodeficiency virus (HIV) which causes AIDS is far more common in the world than previously thought. UNAIDS and WHO estimate that over 30 million people are living with HIV infection at the end of 1997. That is one in every 100 adults in the sexually active ages of 15 to 49 worldwide.[7]

Like its predecessor, the report failed to inspire. The prose still wasn't purple enough to bring the money rolling in, obviously.

We were going to have to get more creative. We took another crack at generating interest and cash six months later, recycling the same numbers but jazzing up the language and the storyline:

> The human immunodeficiency virus (HIV) continues to spread around the world, insinuating itself into communities previously little troubled by the epidemic and strengthening its grip on areas where AIDS is already the leading cause of death in adults . . . The majority of those now living with HIV will die within a decade. These deaths will not be the last; there is worse to come . . .[8]

Cue drum rolls and long shot of storm clouds gathering over the plains.

The story in Africa merited the purple prose. But to find money for other continents, we had to beat things up a bit. When a journalist talks about 'beating it up', they mean making a mountain out of a molehill, making a big, interesting, dramatic story out of something that may actually be rather mundane. There is a huge difference between *making* it up (plain old lying) and *beating* it up. A journalist who beats up a story is a bit like a photographer who takes a picture of a politician from their best angle in soft focus if it is for an election campaign poster, but takes the picture from below, looking up the nostrils under a neon strip light if it is for a tabloid newspaper that supports the other party. The subject matter is the same; the art is in how you present it.

Some people have recently accused UNAIDS epidemiologists of making it up, of deliberately overestimating the epidemic.[9] I never saw any evidence of this. The 'epi' team, as we were known, always did the best job possible with the information and tools then available. As the information and tools improved, the estimates changed, and the agency had the guts to say so. In 1997 the estimates changed upwards. In later years they were more often revised down. UNAIDS publishes its data sources and its methods, and gives reasons for any revision, and that's admirable. The warty truth is that in the early years, the tools and information weren't very good. In 1997 the best we had to work with was a super-simple model in which you decided when the epidemic started, plotted on a graph any information you had about the percentage of adults infected with HIV, and then drew a curve through all your bits of information, a sort of glorified 'join-the-dots' exercise.[10] This was fine as long as you had lots of dots. But in many countries, including countries as big as Nigeria, Brazil and China, we had at best two or three bits of information from just a couple of years. We believed at the time that HIV's epidemic curve would rise like a mountain to a peak, and after a few years come down again. But if you've only got a couple of dots to join, you can't tell where you are on the mountain. Are you on the upward slope or the downward slope? Close to the summit or still in the foothills? It was a bit like trying to draw a picture of Mount Everest when all you have to work from is the position of base camp and photos of one or two spurs of rock, taken during a blizzard. Worse, in fact, because you don't have any idea how high the mountain might be. You could be dealing with Everest, or Mont Blanc, or some local hillock.

The estimates were developed by Number-Cruncher-in-Chief Neff Walker and Epi-boss Bernhard Schwartländer. They sat in Geneva poring over the data, and made endless phone calls on crackly lines to people in distant health ministries who might be able to fill in some of the gaps. They ran and re-ran the models as new information came in. But sometimes, when you don't have the data, you just have to make an executive decision.

Those decisions were often made after midnight and a couple of beers. Neff was unusual even by the standards of the iconoclasts who ricocheted around UNAIDS in its first years. He was a straight talking Texan who chewed tobacco, listened to country and western radio over the internet and went ice skating in his lunch breaks. One of his first executive decisions had been to put a fridge in the office we shared, and I wasn't complaining. Which explains the beers. But the after midnight? As I said, everything at UNAIDS seemed to happen in a last-minute panic. At five o'clock on a Friday evening, you could be pretty sure that one of Executive Director Peter Piot's staff would come crashing in demanding slides for some Monday morning presentation that had been scheduled for months. If we caught sight of them coming down the corridor, Neff or Bernhard or I would mouth to one another, 'Sorry, but Peter wants . . .' Seconds later, they'd crash into our office and say, 'Sorry, but Peter wants . . .' Then they'd swan off for the weekend, leaving us to pull out data, put together slides, make up a presentation.

If you weren't on some impossible deadline, one of your mates was sure to be; often, you'd sit around finding stuff to do, just to show solidarity. Usually, you worked on your own panic. Sometimes, you played computer games. Our favourite was one unearthed by Bernhard, our fearless leader. It was a variation on PacMan where you had to drop condoms onto penises that scooted back and forth across the screen. One night, after being roundly beaten at this game by the much more experienced Bernhard, I walked back into my office to find Neff – the object of our solidarity that evening – sprawled in his chair with his Timberland boots on the desk, cracking open a beer. 'Good day?' I asked. 'Pretty good,' Neff replied. 'We've just saved about a million Nigerians.' He'd just got another tiny scrap of data, made an executive decision and changed the slope of the curve. Hey presto, the estimated number of Nigerians with HIV dropped from over three million to 2.3 million.

The margins of uncertainty in this business were substantial. We could 'save' a million people here or there and still not make any difference to the truth in Africa: HIV was chewing up lives, decimating families, destroying whole communities. When we made colour-coded maps of the HIV epidemic, East and Southern Africa were bright scarlet, and West Africa was a deep red. But this alone did not seem to be enough to spur rich countries or even international organizations to come up with much cash. We needed to make a case that the fire was spreading to other continents – to Latin America, Eastern Europe and above all Asia, home to a quarter of the world's people. But on our coloured maps, these areas stayed cheerily yellow – there just wasn't enough HIV to worry about. How could we make China scarlet? Russia, India, Indonesia? These are the countries we wanted to draw attention to, because their huge populations and booming economies roused the interest of the 'international community', and because we really believed that effective prevention in those countries could save hundreds of thousands of lives.

We sat around, listened to twangy country and western radio, and played with numbers. Inspiration! Don't look at absolute numbers, look at rates of change. The smaller an epidemic is, the easier it is to double it. If Country A goes from one infected person to two, the epidemic has grown by 100 per cent with just one new case. But let's say country B has a much bigger epidemic, a princely 10 infections. If that increases to 15 infections in a year, Country B's epidemic has only grown by 50 per cent. In other words, B has half the growth rate of A, even though they've got five times as many new cases as Country A.

We compared the 1997 estimates with the previous ones, made by WHO three years earlier.[11] Sure enough, the whole palette changed in favour of lower-prevalence countries. Russia went from an estimated 3,000 infections in 1994 (about 0.004 per cent of the 'sexually active' population) to 40,000 three years later. A total of 37,000 more infections, and still just one adult in 2,000 infected, true, but a whopping 11,500 per cent increase in four years. Russia was now bright scarlet on our map. Meanwhile, the proportion of Kenyans infected with HIV increased from 8 per cent to 12 per cent. That added 600,000 new infections and left one adult in eight with HIV, but it was a mere 50 per cent increase. Kenya stayed a delicate pink in Figure 1 of the UNAIDS Global Report.* Moral of the story: always be deeply suspicious when you hear phrases like 'one of the world's fastest-growing epidemics'. It's the first sign of a beat-up.

Win some, lose some

We had another trick too, one that would come back to haunt us. By 1998, we knew the sparks of HIV in Asia, Latin America and Eastern Europe were likely to set ablaze epidemics among drug injectors and gay men, as well as in the sex trade. That put over 100 million people at high risk in Asia alone – the equivalent of about a third of the entire adult population of Africa. There are probably millions more in Europe, North America, Latin America and North Africa. Some of these people would pass on infections to wives or other sex partners. But HIV wasn't going to rage through the billions in the 'general population', and we knew it. All the evidence suggested that people in our

* You'll find a copy of the map on http://wisdomofwhores.com/references.

newly scarlet countries don't weave sexual networks that would allow that to happen.

All the evidence also suggested that governments don't like spending money on sex workers, gay men or drug addicts. Not donor countries, and not scarlet countries either. There are no votes in being nice to a drug addict.

We had to find a way to translate the truth into something that governments might care about. We came up with two options: money and babies. Politicians could save both, we argued, by paying more attention to HIV while it was still low. It was a long shot. In most of the world the money argument will backfire and the babies argument isn't true.

First the money. HIV prevention is relatively cheap. For the price of a condom or a sterile needle today, you can save yourself several thousand dollars in health systems costs caring for an AIDS patient ten years from now. But it doesn't follow that it is a good investment for a prime minister.

As a lobbying tool, the 'buy needles now, save on hospital costs later' argument rests on the assumption that politicians give a damn about what happens 'later'. If later is after the next election, then it doesn't wash. If you do something cheap but unpopular now, you'll get voted out of office. It is no consolation that some opposition government will save money in the future.

So in fact, the money argument often isn't enough to make politicians do nice things for junkies. How about the babies argument, then? Politicians are always happy to do nice things for innocent women and babies. Perhaps if we could show that doing nice things for injectors would protect innocent women and babies . . .?

We argued quite truthfully that men who inject, men who have sex with one another and men who buy sex are likely to pass HIV on to their innocent wives. And then came the sleight of hand. Once innocent wives were infected, we implied, HIV would blaze through the 'general population'.

Here's an example from the 1998 UNAIDS report:

> [In India] the virus is firmly embedded in the general population, among women whose only risk behaviour is having sex with their own husbands. In a study of nearly 400 women attending STD clinics in Pune, 93% were married and 91% had never had sex with anyone but their husband. All of these women were infected with a sexually transmitted disease, and a shocking 13.6% of them tested positive for HIV.[12]

The figure came from one small study of women with sexually transmitted infections. Most of these women were probably married to men who visit prostitutes, already a minority in India. They didn't represent the 'general population' in any way. But the way I wrote it pretty much implied that HIV was raging through the faithful wives of India.

We weren't making anything up. But once we got the numbers, we were certainly presenting them in their worst light. We did it consciously. I think all of us at that time thought that the beat-ups were more than justified, they were necessary. We were pretty certain that neither donors nor governments would care about HIV unless we could show that it threatened the 'general population'. We needed them to care if we were going to get more money out of them, and we needed more money to pay for better prevention. But we also knew that gloom and doom are double-edged swords. Why should people give you more money if you are failing so badly with the money you already have? So as I churned out report after report – one in 1997, two in 1998, one in 1999 and another two to start the new century* – I found myself trying to walk the tightrope between 'It just gets worse and worse' and 'We know how to stop this thing.' The bad news was easy to find; it piled up on our desks with every new study among drug injectors in Burma or teenagers in South Africa, with every new estimate of the millions infected with HIV and the hundreds of millions at risk. But the good news, the success stories that might encourage governments to give us money because they thought we might use it to prevent HIV effectively – those were harder to come by.

I wandered the corridors asking colleagues for good news. I rehashed some stuff from Thailand, Uganda and Senegal, the sum total of developing countries that could demonstrate at the time that their prevention efforts had put a lid on their HIV epidemics. Roeland Monasch, who worked in the WHO youth programme and was part of the Epi department's late-night solidarity team, had some good data, so youth was in. My favourite Geneva dinner companion was Isabelle de Vincenzi, an epidemiologist who headed one of the most important early studies of sexual transmission of HIV in Europe. In her kitchen I learned how to make a 'clafoutis' fruit pie, and how to prevent HIV infection from mother to child – an exciting new science at the time. So there was another good news chapter for the report.

The report-writing was fuelled by hot coffee, cold beer and the occasional foray into the subsidized canteen for solid fuel. On one such foray, just as we were wrapping up the report, I found myself sitting opposite a certain Andrew Ball, an Australian doctor with hypnotic green eyes. Andrew specializes in addic-

* UNAIDS issues annual Epidemic Updates ahead of World AIDS Day, 1 December. Since 1998 it has also published Global Reports on HIV to coincide with the world AIDS conferences held every two years.

tion; he was working in the WHO's programme on substance abuse at the time. Substance abuse? Ah yes, drugs. Amazingly, no one had mentioned drugs on my wanderings through the corridors of UNAIDS, even though we were beginning to see signs that injection was propelling HIV through many parts of the former Soviet Union, not to mention Asia. I don't think the omission was deliberate; it was just that the sexual epidemic in Africa was so overwhelming that it was knocking other things off the radar screen. I looked into Andrew's eyes, in a 'tell-me-more' sort of way.

When the report went to the higher-ups two days later it included five pages on drug injection. And it explicitly endorsed politically fraught policies of 'harm reduction', including programmes that provide access to sterile needles for injectors. In those chaotic early days, when the UNAIDS Global Report on HIV was nursed by epidemiologists in consultation with sundry mates encountered in the canteen or at the dinner table, the only real criterion for inclusion was good data. We didn't care whether needle exchange was politically acceptable or not. We cared whether there was strong evidence from well-designed studies that it worked. Since there was, we argued for more sterile needle programmes. Head honcho Peter Piot read the draft report; he is a good scientist and he allowed the evidence to stand. So needle exchange was on the international agenda, thanks to a chance encounter with a cute boy in the canteen.

Those haphazard days are long gone. Six writers worked full time for six months to put together the 2006 UNAIDS biennial report, and swarms of others helped with analysis, graphics and the like. It cost somewhere between US$1.3 million and US$1.7 million, according to Geneva gossip.[13] At 640 pages, the 2006 report weighed in at 2.1 kilos.* And no wonder – the report has to go through the mandate police of ten different UN agencies, all of them wanting to be sure that their institutional wares are nicely laid out for all to see.† It seems laughable now that I banged out the sixty-three pages of that first UNAIDS biennial report in less than a month, at a cost of US$8,900. Our expectations weren't that high; at best we hoped the report might get a bit of attention and bring in a bit of cash.

Blow me, it worked.

That 1998 report struck the lobbyist's jackpot – the front page of the *New York Times*. PARTS OF AFRICA SHOWING H.I.V. IN 1 IN 4 ADULTS proclaimed the headline.[14] Suddenly, the world was waking up to the fact that whole nations were being ripped apart by a disease that could have been prevented with a piece of latex costing a few pennies. *New York Times* reporter Lawrence Altman concluded that the report 'paints the gloomiest picture of the H.I.V. epidemic since it was first recognized in 1981'. But he picked up on our success stories, too – Uganda, Thailand and Senegal – and quoted Peter Piot as saying that we needed more money, and more political courage. The money started to roll in.

The first institution to be shaken out of its torpor was the World Bank. In 1997 the Bank's analysts had made a good case for paying more attention to AIDS[15] and a few enthusiasts in country offices had pushed through AIDS loans, but the big shots who held the purse strings dozed on, uninterested. A lonely

* Pity the cleaning staff at Toronto airport. They had to manhandle stacks of copies ditched by participants at the 2006 Toronto AIDS conference, who found that the document put them over the airline weight restrictions.

† As the AIDS funding honeypot began to swell, another four UN agencies discovered that HIV was part of their core mandate and joined the original six 'Co-Sponsors' – the International Labour Organization, the World Food Programme, the United Nations High Commissioner for Refugees and the United Nations Office on Drugs and Crime.

AIDS enthusiast at Head Office seized on our '30 million infected' estimates. She herded us over to Washington, a troop of epi-nerds in from Geneva, London, Kampala, Harare, to bang the drum for HIV spending. Could we wake up the big shots from the Bank's Africa division, corralled into the meeting under sufferance?

Yet again, we trotted out the appalling numbers. In parts of Africa, death rates among young adults had tripled since the pre-AIDS era; even in rural areas HIV caused seven out of ten deaths in adults under sixty. Life expectancy was wilting back to levels not seen since the 1940s . . .

Life expectancy? That's one of the key measures the World Bank uses to show that it is helping countries to 'develop'. If it gets shorter rather than longer, the Bank can be judged to be doing a bad job. Heads might roll. Jobs-for-life might be lost, perhaps even before a big shot reached fifty, the age at which a World Bank executive can retire on a full, six-figure annual pension. The following month, World Bank President James Wolfensohn told African leaders that AIDS 'needs to be put front and centre'. Soon, the Bank had stumped up an extra US$500 million for AIDS in Africa.[16]

Africa alone didn't seem quite enough to inspire other institutions. But the *New York Times* story also drew attention to our 'rapid growth' countries of Asia and Eastern Europe. 'An alarming trend is the doubling and tripling of transmission rates in some countries in these areas since 1994,' the story ran.

This was picked up by the National Intelligence Council, the US government's official security think tank. The NIC has not traditionally been a hotbed of epidemiological forecasting, but they were casting around for a new role in the post-Cold War era. They started playing with numbers and by 1999, they declared that HIV was a threat to America's national security, not least because the virus was 'spreading quickly in India, Russia, China, and much of the rest of Asia'.[17] The report was a bit hazy about the link between deaths in India and security in the United States, but never mind. There's no better way to open the purse strings than to have a threat to national security up your sleeve.

The US Congress immediately asked the development agency USAID to cost a global 'Battle Plan' against AIDS, and USAID asked for our help. In 2000 USAID informed Congress that between US$3 and US$5 billion would be needed each year just for sub-Saharan Africa.[18] By 2001, Jeffrey Sachs, a much-quoted guru of international economics, was demanding US$7.5 billion a year for AIDS, and UNAIDS raised the bidding to US$9 billion.[19]

The demands were getting bolder because money was indeed coming in. The pot of money available for AIDS languished around US$300 million a year in 1996. It sailed past the billion-dollar mark three years later, and continued to grow. I don't think any of us dreamed back then quite how much it would swell – to US$10 billion by 2007.[20]

A big chunk of that money comes from US taxpayers. In 2000 they were ploughing about US$233 million a year into HIV programmes in developing countries, with the bulk of that quite sensibly going to Africa. All other rich countries put together handed out another US$680 million. African governments paid up US$85 million, while governments in Latin America, Asia and Eastern Europe shelled out around one billion.[21] Almost all of this was destined to finance prevention – providing medicine for people with HIV was seen as tantamount to throwing money into a bottomless pit, and most donors just didn't do it.

Three years later, boom! Bush announced that he would spend US$15 billion on AIDS in the developing world over the next five years, over eight billion of it on treatment. In May

2007 he told Congress that he was going to ask for another US$30 billion for five years beginning in 2008.

The impetus for this came from über-Republican senator Jesse Helms. That would be the same Jesse Helms who in 1995 tried to squash funding for AIDS programmes at home because HIV is spread by the 'deliberate, disgusting, revolting conduct' of gays.[22] His eyes were opened during a conference organized by Franklin Graham (son of televangelist Billy), who added his spin to ours. Graham pitched HIV not as a disease of the wicked but as an affliction of the innocent. AIDS was no longer about promiscuous queens in rich countries, it was about loyal wives in dirt-poor countries. Their helpless, wide-eyed infants were suffering, too. These are good, God-fearing people who could be helped by Christian charities, if only Christian charities had access to more money to help them with. Helms helped orchestrate a formidable lobby in Congress, and suddenly the world was looking at hallucinatory amounts of cash for HIV prevention and care.

AIDS was a growth industry.

Cookbooks

In 1998, writing a 'Report on the Global AIDS Epidemic' was not a full-time job for one person, let alone for a team of half a dozen. I did all sorts of jobs for UNAIDS and other members of the AIDS mafia in between reports. Want three volumes on strategic planning for AIDS programmes? No problem. Some guidelines for HIV testing in clinics for pregnant women? Sure thing. A technical report on the interaction between sexually transmitted infections, HIV and infertility? Consider it done. These dry technical jobs came as a relief after the manipulative gymnastics of the Global Reports. I got to plug into the brains of people who knew far more than I did, and it wasn't a bad way to learn which sliver of the AIDS cake interested me most.

It turned out what interested me most was HIV surveillance, the source of a lot of the numbers we were throwing around in the Global Reports. How did we know how many people are infected with HIV, where they were, who they had sex with? How did we know where the disease might go next? How could I make statements like the one I just made: 'HIV wasn't going to rage through whole populations, and we knew it'? Much of this information was supposed to come from HIV surveillance systems, but they weren't up to the job.

A lot of disease surveillance is plain old counting. Health facilities keep records of the diseases they see, they report cases up through a government information system, and some gnome counts them up. Not exactly captivating work for a girl who was expecting a career in sex and drugs.

As usual, AIDS breaks the mould. Reporting AIDS cases isn't useful if you want to know how HIV is spreading. That's because most people don't show any symptoms of AIDS until they've been infected with HIV for a decade or more. An AIDS case that shows up in 2008 was probably a new infection around 1998. AIDS cases tell you nothing about how the virus is spreading right now, which is what you need to know to plan sensible HIV prevention programmes. Making HIV prevention decisions based on reported AIDS cases is like looking at 1998 stock prices to decide which shares you should buy in 2008. Why would you do that if you could get more recent information?

HIV testing gives us more recent information. In 1989 WHO recommended doing HIV tests on left-over blood that had originally been taken from pregnant women to test for syphilis.[23] It was assumed that pregnant women represented the sexually active population. Testing their blood for HIV anonymously would be a convenient way to get an idea of rates of HIV infection in all adults. This was fair enough in many African countries, where HIV spreads mostly in sex between men and women. But in the rest of the world, it was daft. In Europe, the Americas, Asia and North Africa, the people most likely to become infected with HIV are drug injectors, most of whom are men. Or men who have sex with one another. Or men who buy sex from women. In other words, the people most likely to be infected are men, and they are not showing up at clinics for pregnant women.

The women most likely to have HIV often don't show up at clinics for pregnant women, either. Across the globe, sex workers have higher HIV than other women. They also use lots of contraceptives (because pregnant prostitutes don't get many clients). When sex workers do get pregnant, they tend to avoid the government pregnancy clinics where blood samples for surveillance are taken. Those clinics are for 'nice' women, they are made to feel. In most of the world, then, testing pregnant women for HIV doesn't tell us much about the epidemic.

Even in Africa, the system was showing cracks by the mid-1990s. The alarm was raised in the early days of UNAIDS by the Ugandan AIDS programme, which had far and away the strongest surveillance system on the continent. Programme staff had noticed that in several clinics, HIV rates were starting to fall. Hurrah! said some people. HIV programmes must be working, because fewer people are getting infected. Not so fast! said others. HIV may be falling because fewer people are getting infected. But HIV prevalence can fall even if more people get infected this year than last. It depends what happens to death. If there are more deaths than new infections, the number of people living with HIV will fall even though the number newly infected is rising. So the fall in HIV rates could be bad news (more funerals), rather than good news (fewer new infections).

In the creaky UN system, it can take rather a long while to go from alarm bells to diagnosis. But by early 1997, UNAIDS prevention guru Michel Caraël had marshalled the troops to a meeting in Nairobi to try and figure out what was going on. Michel was a chain-smoking Belgian anthropologist who had become a specialist in the measurement of sexual behaviour. He suspected that the only way to make sense of HIV surveillance data was to overlay it with information on trends in unprotected sex with multiple partners. If risky sex was falling, then the lower HIV could probably be chalked up to good prevention programmes. If there was no change in unprotected sex, then any fall in HIV probably just meant infected women were falling out of the pool being tested because they were dead or not getting pregnant any more.

I knew a bit about measuring sexual behaviour from my studies of demography, so Michel invited me to the meeting as scribe. Colleagues from Uganda, Zambia and Zimbabwe fished around for data that could help make sense of what was happening. We argued loudly about what the results meant, shocking the waiters at Nairobi's Norfolk Hotel, a genteel relic from an earlier age, an age of British rule when guests did not discuss condoms in booming French accents. When Michel started telling stories of drinking foreskin soup at a circumcision ceremony in Congo, the waiters were faint with relief that we had insisted on sitting outside, away from the crusty old colonial couples.

After much booming and hilarity, we concluded two things. Firstly, what more or less worked for Africa wouldn't work anywhere else. In most of the world we would have to track the spread of HIV in the places it was most likely to be spreading –

the brothels, the gay bars, the heroin shooting galleries. Secondly, we couldn't get away with just testing blood for HIV. If we wanted to be able to explain changes in HIV rates, we had to know whether behaviour was changing as well. Measuring behaviour should also help us get a handle on what was likely to happen next in the epidemic. If a lot of drug injectors are sharing needles *and* selling sex, we can expect HIV to rise among both injectors and prostitutes. But if they also use condoms with their clients, we don't have to worry too much about men who buy sex, let alone innocent women and babies. HIV surveillance was more interesting than just counting case reports in health centre databases.

In the state-of-AIDS-in-the-world reports, we spent a lot of time nagging countries to look at the facts and act on them. But you can only nag people to stick to the facts if you are confident you know what the facts are. HIV was already one of the best-documented infectious diseases in history, which does not say much for the others. But a lot of our facts came from special studies, from academic work, from systems that broke down when a researcher was pregnant, a grant was cancelled, or a global AIDS programme was dissolved. The Nairobi meeting got me thinking how much better we should be doing with routine government surveillance systems. If we started doing surveillance better, focusing on the right groups and tracking behaviour as well as disease, we'd have a much better idea of how HIV was spreading. But we'd also have a much better idea of what we should be doing about it. And because we could track changes in behaviour, we'd have a way to measure whether what we were doing was making any difference.

As the beat-ups started to work and the money began to come in, it became more important to be able to show that we were actually doing something with the money, and to see if what we did made any difference. And it seemed important, too, to get better data from the groups that politicians wanted to close their eyes to.

Between 1997 and 2001, I spent progressively more of my consultancy hours, weeks, months on surveillance. I morphed from 'writer' to 'expert' – a fully fledged member of the HIV surveillance mafia.* We had meetings in Nairobi and Geneva and Washington, we sat on Alpine lakefronts, Thai beaches and dreary university campuses in North Carolina. I stacked up airmiles – one summer I spent more hours flying British Airways than my cousin Michael did, and he was a pilot with the airline. Instead of just listening at meetings, I was invited to speak. I started writing papers in scientific journals and chapters in text books. And I wrote a lot of 'How-to' manuals for UNAIDS and its various mafia friends – the cookbooks that were supposed to help countries whip up a better surveillance system.

The trouble with UN cookbooks is that they're supposed to be as useful in Afghanistan as in Zimbabwe. Often, that means deleting anything specific. I can't talk about transgender prostitutes because in some places they're transvestite not transgender. I can't talk about heroin injection because in some places the main drug injected is cocaine. I can't talk about extramarital sex, because what we are really interested in is multiple partnerships, and in polygamous countries you can have multiple partners without having extramarital sex. And so it goes on. In the end,

* The surveillance mafia was made up of people from lots of different institutions – UNAIDS and WHO of course, but also UNICEF, the US Center for Disease Control, a number of research organizations, some people hijacked from national AIDS programmes and sundry others. I worked on contract for several of them, and most of the documents we produced groaned under the weight of all their different institutional logos. It was a rare example of a lot of people working together because we really wanted to, rather than because some UN steering committee had decided we should want to.

it sometimes felt like having to write a cookbook without being able to mention beef or salmon or grilling or oven temperature: Take some food. Cook until done.

It was fun, though, and the money wasn't bad, at least compared to journalism. But there were downsides. For one thing, consultancy work is inherently unsatisfying. Consultants write manuals and pontificate about how things should work, but they rarely actually try to make things work. After delivering their wisdom they check out of the hotel and drive to the airport. More importantly, consultancy is not that great for one's personal life. I rarely saw the boy with the giant, generously rent-free house, even though we'd married in 1998, the week that the UNAIDS beat-up report hit the front page of the *New York Times*. David was running the Reuters news agency bureau in Nairobi, covering all of East Africa. Just as I arrived back with a nice, tranquil three weeks to spend writing a report at home, the Ethiopians would launch another assault on the Eritreans, or vice versa, and he'd take the flak jacket off the back of the door and fly off to the front lines. He'd come back shattered and all turned in on himself, keen to clear his head on the golf course or fug it up in a bar. By the time he was ready to talk again I'd be on my way to the airport for some job in Botswana or Geneva or Cambodia.

I wanted a 'real' job, one that earned me fewer air-miles, one that allowed me to spend less time telling other people how to do surveillance and more time doing it. The AIDS industry was beginning to boom, and the organizations I had worked for as a consultant were looking for staff. Africa was awash with jobs, but I could have worked just about anywhere.

Anywhere except Singapore, perhaps. And of course Reuters posted David to Singapore to give him a rest from nights in military fox-holes in Ethiopia, days in refugee rat-holes in

Congo, years of adrenalin, feigned fearlessness, courage, burnout. Good for a resting journalist, but useless for a hired-gun HIV epidemiologist. The hyper-organized Asian city-state's public health service was well stocked with indigenous talent, and there was virtually no HIV. I looked around. The Indonesian capital Jakarta was only an hour from Singapore – that seemed like a reasonable commute after the nine-hour Nairobi to Geneva run. Better yet, the US government had just slapped US$14 million on the table for an HIV prevention programme in Indonesia, and the contract to run it had gone to Family Health International, employers of some of my favourite members of the HIV surveillance mafia.

FHI is a non-profit group based in Washington, but it was their Bangkok office that had more or less set the global standard for tracking risk behaviour among sex workers and clients. Steve Mills and Tobi Saidel, who were by then blazing trails in surveillance with transgender prostitutes, drug injectors and gay men all over south and south-east Asia, were hard-working, smart and fun to be with. Did they need anyone to help out on surveillance in Indonesia, I asked casually at the start of 2001? As it happens, yes. It wasn't a stay-at-home job – I'd be expected to spend up to half my time helping out in other countries in the region. But it was an improvement on the two-weeks-here-three-days-there model that had begun to enslave me in Nairobi. It was as close to David as I was likely to get, and there were jobs opening up soon in the Reuters bureau in Jakarta, so perhaps he'd be able to join me. Above all it was a real job, working with real governments, doing real surveillance.

There was one little problem. I was being sold to the Indonesians as an HIV surveillance expert, and on paper I looked OK. But I'd never done any surveillance. I'd never collected a blood sample, never been in a laboratory, nor drawn up a statistical sample frame, nor coded a questionnaire. I'd written manuals about how to do these things, but I had never actually done them. On top of that, I knew that the landscapes of sex and drugs in Asia had more in common with what was going on in industrialized countries, in Eastern Europe, Latin America and North Africa than with the epidemic in sub-Saharan Africa. Yet the AIDS mafia had been hypnotized by Africa, where 'men infecting women' plus 'women infecting men' equalled mind-bending infection rates. I knew a lot less about the more complex equations of Asia, where a man might infect a man who infects a drug injector who infects a sex worker who infects a client who infects a wife.

Like so many 'experts' the world over, I was going to have to fake it.

2

Landscapes of Desire

I felt all wobbly inside when I touched down in Indonesia in 2001. It was one thing to pontificate about global trends in HIV infection, quite another to try to zoom in on a single red light district, a gay scene, a clutch of injectors. I wasn't sure what to expect of the job, of my colleagues, of a 'regular' life. But I was excited, too. I had loved the riotous diversity of Indonesia when I worked there as a journalist ten years earlier. Since then, Indonesians had sent a thirty-year-old dictatorship to the scrap heap and grown more riotous. My own position had changed almost as radically. As a journalist I'd spent my time criticizing the Indonesian government. Now, I'd be working for it.

I was supposed to help the government improve its understanding of how and why HIV was spreading in Indonesia. That was the nerdy part of the job, building national surveillance systems that would generate good information year after year, and that could evolve with the epidemic over time. It would be fun, but it wasn't enough for me. There was no point getting the information unless we used it to improve HIV programmes. That was probably going to mean persuading the government to do nice things for drug injectors, sex workers and gay men. It was going to mean finding money to pay for those nice things, maybe even re-evaluating how the US money we were working with would be spent. I wanted to understand the landscapes in which people had sex and took drugs, but I was also hoping to meddle in reshaping those landscapes.

In my days as a journalist I had travelled the length of Indonesia's 17,700 islands, the same distance as London to Tehran. I knew a bit about the clove trade in Indonesia, the stock market, the pest management system in rice crops. I knew about sporting successes and banking bailouts, about 'mass movements' and mass graves – the last two both political tools of President Suharto's now defunct government. I'd listened to friends who had lived through the economic bust of the late 1990s fret as the gap between rich and poor grew ever wider and the voice of Islam ever more strident. There are more Indonesians on this planet than people of any other nationality with the exception of Chinese, Indians and Americans. Indonesia is home to more Muslims than any other country, but it is a secular state with five officially sanctioned religions. It has one official language and 736 unofficial ones. Nothing in this magnificent, chaotic nation is simple. That includes HIV.

The Indonesian government had been testing sex workers for HIV since 1991. American and Indonesian researchers had begun more elaborate studies at the same time. Colleagues at FHI and the University of Indonesia had surveyed risk behaviour among sex workers and clients every year since 1996. But the results of all this measurement didn't seem to make much sense.

Like most South-east Asian countries, Indonesia has a big sex trade, and all the surveys suggested that condom use was very low and sexually transmitted infections (STIs) were very high. A recipe for HIV, surely? As far back as 1992, an American researcher called Mike Linnan published an apocalyptic report, 'AIDS in Indonesia: the Coming Storm', claiming that a tide of infection was about to engulf the country. Linnan and others banged the warning drum about a massive sexual epidemic. In 1994 they predicted that if nothing was done, the low-condom + high STI + large sex trade equation would produce half a million HIV infections by 1998. The World Bank got so excited that in 1996 it lent the Indonesian government nearly US$25 million to deal with the problem. Government employees busily dealt with it by setting up NGOs and channelling money to themselves.* The mismanagement was so gross that the project was eventually cancelled, and all but one of the forty-one NGOs that had been set up in one of the two provinces where the project worked immediately disappeared.[1]

Other HIV prevention projects financed by the US and the Australians since the mid-1990s managed to stay on their feet. They ran a baffling poster campaign showing people playing pool and the slogan 'AIDS, wait a second.' They provided some treatment for sexually transmitted infections, gave out condoms, got prostitutes to talk to one another about HIV. But they didn't seem to be making much difference. When I arrived in Indonesia in 2001 to help out with the second wave of the US-financed programme, Linnan's storm clouds were all rumbling along: there was still a large sex industry, nine out of ten clients still didn't bother with condoms, the red light districts were still awash in STIs. But the storm of heterosexual HIV that had been forecast for so long had not hit. In fact the only clear indication that HIV was rising came from drug injectors.

* Government-linked 'NGOs' are so common in Indonesia that they have a nickname: 'Red Plates' after the colour of the numberplates issued to official cars.

Going through my notes late at night on the day I arrived in Jakarta I was bemused, but there was no point losing sleep over it. The very next day there was a big national meeting on HIV surveillance in Indonesia; doubtless someone would explain the mystery.

They didn't. In fact, they compounded it. At the meeting a distinguished retired doctor raced through the numbers to explain how the government had arrived at an estimate of 100,000 HIV infections nationwide for 2001. There are 150,000 prostitutes, at 3 per cent infection, so we'll round that up to 5,000 infected prostitutes (he used the term WTS '*wanita tuna susila*', women without morals). Let's say each prostitute infects five clients. Another 25,000 infections. Then there are maybe 250,000 drug injectors, at 25 per cent infection: 62,500. We'll add in another 10,000 to cover housewives and stuff. So total: 102,500.[2] High science.

I was none the wiser. The records show that over 6,000 sex workers had been tested for HIV nationwide the previous year and just fifty-four were infected – less than 1 per cent. It seemed a bit extreme to use triple that rate in the national estimates. Over half the estimated infections were in drug injectors. That may well have been true, but it was based on pretty flimsy evidence. In 2000 only 157 injectors were tested for HIV, all in the same treatment centre in Jakarta; sixty-four of them were infected. Other groups didn't appear in the estimates at all, simply because no one was testing them. There was not a shred of information about gay men, for example. And none of the girls I'd seen strutting their stuff along the railway tracks just the previous evening seemed to be on the radar screen. That's because despite the colourful parade of high heels and short skirts, bright lipstick and dark looks, none of them was actually a girl. They were 'waria', and no one had bothered to measure HIV infection in waria since 1997. No data equals no problem.

A waria is a smush-up of 'woman' (*WAnita*) and 'man' (*pRIA*). We tend to translate this as 'transgender sex worker', but the term seems flaccid in the face of this throbbing subculture of biological males who live as women and sell sex to men. The very first indications that HIV was circulating in Indonesia came from studies among waria. The last time anyone had taken the trouble to find out, in 1997, 6 per cent of waria in Jakarta were infected with the virus.[3] Now we were ignoring them, along with rent boys and gay men. In fact, according to the national estimates, no one in Indonesia got infected with HIV during anal sex.

I asked the learned doctor about possible infections in anal sex between men and got the brush-off. 'There aren't enough waria to worry about.' But that wasn't my point. The waria's clients are not wearing high heels and miniskirts. They are in jeans and leather jackets, in civil servants' uniforms or shirts stained with sweat from unloading timber on the docks. There may not be all that many waria, but we knew there was HIV in this group, and we knew that waria didn't infect one another. Their infections come from their 'straight' clients. And there are a lot of clients.

Since the days when transgendered high priests sat in magnificent wooden sailing schooners that plied the waters of the archipelago in the service of the rulers of South Sulawesi, Indonesia has absorbed the concept of men dressing and acting like women. If a man has sex with a waria, it doesn't really count as sex with another man. People don't think too hard about the details – whose appendage goes into which orifice. But HIV is a virus that doesn't care what people think. It only cares what people do. If people have anal sex with lots of partners the virus gets a boost – anal sex opens doors for HIV, no matter what the people who are having it are wearing.

Sex in boxes

I sat in my little office in the grounds of the Ministry of Health, watching rotund civil servants fail to touch their toes in an enthusiastic Friday-morning exercise ritual. Inspired by the fluency of two of the three other foreigners in the FHI office, I dredged up my Indonesian and went around meeting my new colleagues. There was the venerable Arwati Soepanto, a diminutive retired civil servant with a stern grey bun and a surprising giggle, who had an almost preternatural ability to coax action out of the sclerotic government system. She doled out quiet wisdom to the likes of Made Setiawan, a gangly Balinese research student with hair to his waist. Made set up one of Indonesia's first programmes to help drug injectors, and his low opinion of the government was not improved when local officials persecuted the NGO's staff. There was Ciptasari Prabawanti, who dressed with Muslim severity and hung out with prostitutes whose condom use she aimed to increase. Even the office janitor Jumiran was surprising – a middle-school dropout, he cheerfully emptied bins and made tea, then stayed late in the evenings, surreptitiously teaching himself software skills and chatting with me in English that he kept hidden from the rest of the staff for fear they would tease him for getting above himself.

Yes, there was plenty that was surprising. Most epidemiologists would rather avoid surprises – we prefer things to be logical. In Geneva, we had spent a lot of time trying to fit the HIV epidemic into an organizational chart of boxes and arrows. People are either in the sex worker box, for example, or the client box or the wife box. But Indonesia was never a country to fit into neat boxes of any sort. I had a hunch that the sex boxes would not be an exception.

Every surveillance cookbook I'd worked on started off with a health warning: don't launch into surveillance without doing 'qualitative research' first. None of them actually had much to say about what qualitative research was, why you should do it, or how. But essentially, qualitative research is the step between 'hunch' and 'epidemiological study'. It comes all wrapped up in jargon: key informant interviews, focus group discussions, semi-structured questionnaires. But it boils down to picking the brains of a handful of people who know more about a subject than you do. You try to find out how to reach the people you want to reach in your study, what sort of questions you should be asking them, what sort of language you should be using. If you do your qualitative research right, you'd know not to ask Indonesian rent boys about 'sexual orientation' because most of them don't understand sociology-speak. You'd know that you shouldn't do research in brothels on Saturday nights because the owner is likely to be there and he doesn't want his girls disturbed. You'd know all sorts of things that would save you time and energy later on, when you draw up your questionnaires, take them out to tens of thousands of people, collect your surveillance data and try to understand what it means.

A lot of epidemiologists are snobby about qualitative research: it's for sociologists and other 'soft' scientists. If you do it by the book it is time-consuming and expensive, and it produces information that can be dismissed with a wave of the hand: interesting, perhaps, but not statistically significant. Epidemiologists would much rather get stuck straight into the serious work of counting things.

I had at least one foot in the 'qualitative research is for

pussies' camp when I arrived in Indonesia. I could see that Indonesia's sexual landscapes are about as varied as its islands, its cultures, its languages. But in HIV surveillance we already had our cookbooks to hand, with our model questionnaires. We knew what questions we needed to answer – questions that would allow us to spit out indicators prefabricated in Geneva and Washington. And we'd been doing surveillance among sex workers and client groups in Indonesia for five years already. Couldn't we get on with counting things?

Not among waria, certainly. There wasn't even a model questionnaire for waria in the international surveillance cookbooks, and we knew almost nothing about their apparently heterosexual clients. Some 'qualitative research' was going to be inescapable. We spent months setting up a study with the University of Indonesia, and we did it by the book, focus group discussions and all. It took for ever, but it was worth it. One of the first people we spoke to was Fuad, a twenty-one-year-old lad who occasionally worked as a truck driver's assistant and who bought sex from waria. Fuad's girlfriend lived in Bandung, a university town in the cool hills east of Jakarta. Because his truck work was intermittent, he occasionally supplemented his income by giving blow-jobs or selling anal sex to men who cruised in one of Jakarta's few parks, outside the Finance Ministry beneath the bulging thighs of the monumental, bare-chested Papuan who was symbolically breaking free of the shackles of Dutch colonization. Sex with men was just a cash thing; Fuad was straight. To remind himself of that, he might occasionally want someone to give him a blow-job. But that's not something you can ask of a 'nice girl'; Fuad shared a common perception that oral sex is insulting to women, including to female sex workers. So he went to a waria, also known less politely as *banci* (pronounced banchee).

'If I go to a banci, well, it's that I'm thinking of my girlfriend,' Fuad told our research team. 'I'm 100 per cent into women. Don't think that because I go to a banci I'm a fag. I'm not into that at all.'

Fuad's girlfriend was doubtless a nice girl. She also worked the streets of Bandung at night. So here we have a self-proclaimed heterosexual guy who has unpaid sex with a woman who sells sex to other men, while himself also selling sex to other men and buying it from transgendered sex workers. He pushed a lot of the 'high risk' buttons for HIV infection, yet he wasn't a female sex worker, a client, a drug injector, a gay man or a student. He didn't fit into a single one of our questionnaire boxes.

The truth is, real people don't have sex in boxes.

Fuad made me realize that counting numbers of partners and tracking condom use was not enough. If I was to make any sense of the 'hard science', I was going to have to do a lot more to understand the landscapes of sex and drugs – who buys from whom, who else they have sex with, where they cruise, how they get high. It was to be a long journey, zigzagging back and forth between 'hard' and 'soft' research. But I also realized that you could get a lot of the information you needed without plodding through the sociologist's toolkit. With the exception of Fuad, most of my best guides were not respondents to surveys. They were the men, women and waria I chatted to in bars or brothels, on street corners or in cinemas, in offices and salons, in police stations, rehab centres and conference halls.

Tika, for example, a waria who sells sex along the train tracks in East Jakarta. Tika is not as complicated as Fuad, thank God. Like most waria, she only has sex with straight men.

'Sex with a homo? No thanks!' said Tika. 'Homos' want to have sex with men, she explained, while waria want to be women. In fact, they *are* women, in their heads and in society. This answered one of the questions that worried at me in my first interactions with waria. In Indonesian, people commonly address one another using family terms: Bapak (father) and Ibu (mother) if the person you are talking to is older than you, Mas (brother) and Mbak (sister) if they're your age. Should I call Tika Mas or Mbak? I fudged, using the neutral term 'anda' (you), but it felt really clumsy. As Tika told us about her 'husband' of fifteen years, it became clear to me that she was a she, regardless of what lurked beneath her skirt. Tika cooks for her husband, washes his clothes, cleans the house. Feminists hold your breath; for Tika that life is the apogee of all that it means to be a Real Woman. Having said that, she still swans off to sell sex on the street whenever she feels like it. 'It's where we see our friends, it's how we catch up on the gossip.' Who's got a sale on stockings, where to get cheap silicone injections, is that cute new policeman any good at volleyball? (they'll find out at the weekly police vs waria volleyball match behind the Grand Melia hotel). Waiting for clients on a street corner is to a waria what going to a mall is to the average Jakarta teenager.

Is being a waria synonymous with selling sex on the streets? If yes, then all waria were probably at high risk for HIV. If no, then we needed to figure out who was selling sex and who wasn't, so that we could focus our HIV prevention efforts on those who needed the services most. To learn more, I went to have a quiet chat with Lenny Sugiharto, a doyenne of Jakarta's waria world, at her hair salon in a neat, working-class district of Jakarta. I got lost in the maze of alleys around her place, and was taken there by a man in a string vest and flip-flops who was bouncing a baby on his shoulder. The local community seemed quite happily to have absorbed a group of men living as women in their midst.

At Lenny's salon, behind careful lace curtains, I met Nancy, head of Jakarta's well-organized waria network. In its structure, the network mirrors the municipal government – a head and vice-head for each of the city's five major districts, and then an overall honcho. Compared with the Jakarta administration, however, the waria network operates at a stratospherically higher level of efficiency. That came in handy when we eventually set up the surveillance system for waria – the district health officers could contact their 'counterparts' in the waria structure, who would set everything up for them.

Lenny was busy, so I put my question to Nancy. Do all waria sell sex? 'It's not like we *have* to sell sex,' she replied. 'It's just that cruising is an important social activity for us. And, of course, we have super-charged libidos.' In her polyester lilac trouser-suit, amethyst necklace and matching *jilbab* (the Islamic headscarf that was rapidly becoming de rigueur among Indonesian ladies of a certain age), Jakarta's Head Waria would fit right into a Women's Institute meeting. For the last eighteen years, she's been employed by Jakarta's Department of Social Affairs to teach hairdressing and catering skills to fellow waria as a way of enticing them out of sex work. I half expected her to start discussing her favourite recipe for jam tart. But here she was, talking dirty. 'It's less about the money than about the orgasms,' she said. 'Let's face it, we're all human, we've got to get laid.' She told me she made it her business to get laid at least once a week. She'll still take money for it if she can (which gave her catering classes the sheen of 'do as I say, not as I do') but unpaid is fine too. 'What fun is life without orgasms?' she laughed. At the time, Nancy was teetering on the edge of 60.

Nancy was furiously opposed to the new fashion for sex change operations among waria, which she put down, quite simply, to showing off. Not breasts of course – all waria have those these days. But the full-on 'op'. Indonesia's surgeons are not up to scratch in terms of the removal of the penis and testicles and the fashioning of a false vagina, so the destination of choice for sex change operations is Thailand, just a three-hour flight north. Thailand has been trying to rid itself of a reputation for sex tourism; its in-flight magazines now overflow with ads for medical services for visitors instead. Walk past the Starbucks in the towering atrium of Bangkok's Bumrungrad Hospital and take a gleaming escalator to the third floor, and you'll find the International Medical Centre, a United Nations of smiling nurses and interpreters eager to help you figure out how much money you could save by stocking up on treatment here, rather than at your home-town hospital. Lead competitor Bangkok Hospital boasts a website which declares, 'Feel the new sensation of life at Sex Change Clinic Bangkok Hospital' in ten languages. If you don't get the message, the hospital can provide interpreters in a further sixteen.

Both of these hospitals rise out of alleyways crowded with tawdry girlie bars that speak of the earlier tourist industry specialization. Indonesian waria can pile aboard a budget flight to Bangkok and turn a few tricks while waiting for their operation. At around US$6,000, a sex change operation is not exactly a snip. Nancy believes it is a way of signalling to clients that a waria has been a success in a high-paying market like Singapore or in the shadows of the Bois de Boulogne, on the outskirts of Paris. 'They come home and stand on the side of the road, not even wearing any underwear, and flash their new pussies at passing cars. It is just showing off. Money and ego is what it is,' she growls. For all the Women's Institute niceties, I'm suddenly aware that it would be no fun being on the wrong side of Nancy in a roadside negotiation.

Emerge after midnight from Jakarta's stylish Four Seasons

Hotel and you'll see what Nancy means. As you step into the miasma rising off the nearby reservoir you'll be assaulted first by the smell, sour, fetid, heavy with the slough of the millions who scratch out lives alongside the city's waterways. Then come the rats, scuttling purposefully over the moonlit skeletons of daytime food stalls. Finally, as you reach the waste land where miasma meets highway, you'll come across Lydia or Regina, Olive or Baby – perhaps all of them. Most waria go for the classic 'Russian hooker' look – the black PVC skirts stretched tight over the fishnet stockings, plunging into the red patent-leather boots. They'll thrust their butt out to the left, dangle their cute imitation Chanel purse to the right – the pose will show off the new breast implants as well as helping them balance on those vertiginous heels.

There's always a lot of flicking of hairdos and endless public display of lip gloss. There are two types of interaction with the slowly cruising cars that hold the lusty, the unfulfilled or the just plain curious. One is a nonchalant disregard; the 'girls' chat with one another and feign mild annoyance at being interrupted by guys wanting to negotiate for their services. 'I don't need you that badly, so you'd better make it worth my while' is the sales message projected by this crew. The other approach is more brazen – glittering gowns are thrown open as a car crawls up; the full panoply of wares put on display. This sales pitch is favoured by those who have invested their savings in 'the snip'. 'Sooooooo over the top,' sulked Nancy. She herself would never trade orgasms for an operation, she says, but I can't help feeling she's a tiny bit jealous.

Lenny, who was putting on her make-up while Nancy pontificated, laughed. Lenny is not thrilled that the Department of Social Affairs has tossed her, along with all waria, into a box marked 'mentally disabled'. But she had to agree with 'over the top'. Lenny had organized a group that was lobbying for equal rights for waria. She interrupted her face-paint ritual to tell of a recent meeting with a parliamentary subcommittee. 'We're in the national parliament asking to be taken seriously as a community, and I see that two of the girls are missing. I send someone off to look for them and guess what? They're screwing the security guards in the bathroom.' She shook her head in disbelief and went back to her mascara.

Nancy perked up again at the tale. She had reported the miscreants to their *mami*, the long-established cell mothers who oversee the younger waria. 'They beat the shit out of them,' she said, with visible satisfaction. 'No respect, that's the trouble with youngsters these days, no respect.'

I had first met Lenny sitting quietly in her *jilbab* and red lipstick during the first planning meeting for the qualitative research among waria and their clients in 2001. Or at least I thought that was when I first met her. But as we spent more time together and started to gossip about our lives, I learned that we had actually known one another since the late 1980s. We were both in different incarnations then; I was a tearabout journalist and Lenny was Mr Eko, the manager of one of my favourite Jakarta restaurants. In fact, he had helped me organize my farewell dinner when I left Jakarta the first time around. Lenny had a diploma in hotel studies, and had worked in the Mandarin Oriental and other posh hotels in Jakarta and Bali before becoming a restaurant manager. It was a solid career path, but it didn't suit Lenny. She'd turn up to work in her white jacket, assert her authority as Mr Eko all day, then get picked up after work by her 'husband', perching in a ladylike side-saddle on the back of his motorbike. 'There I was, Mr Eko, taking injections,' she waved vaguely at her breasts, 'being picked up from work by my husband like all the ladies. It just didn't add up.' She quit her job and floated into life as a waria, opening a salon, organizing fellow waria, and eventually becoming a stalwart of our research team.

We followed up the qualitative research with a solid survey of syphilis and HIV among waria in 2002. The outcome set Lenny on yet another course. When the lab technician first handed me the test results from that survey, with the positives highlighted in red, I was stunned. For the first and so far only time in my career as a researcher, I felt hot, salty water brimming from my eyes. These tests came from the waria that we'd been teasing, discussing film stars with, playing agony aunt to. One in four of them was infected with a fatal virus, one in four would be dead within a decade, at best, unless we could get them treatment. We'd expected to find some HIV, of course, but nothing like this. When you calculate the number of people you need to include in a study, you have to make assumptions about how much disease you'll find. I'd assumed a worst case scenario of 10 per cent prevalence. Here I was looking at more than twice that. I think I cried as much from the shock of it as anything.*

Lenny had been planning to set up an NGO to provide 'IEC' to waria. That's AIDS industry jargon for 'information, education and communication' – essentially telling people how AIDS is spread and how to avoid it. She had applied to FHI for funding. But as soon as we saw the study results she rewrote the application. It was too late for prevention alone. Waria were going to need doctors, drugs, carers. They also clearly needed treatment for other STIs – close to half of the

* We presented the results to waria community groups with posters that showed 100 waria, one in four of them coloured-in in red. We also presented condom use rather graphically. See http://www.widomofwhores.com/gallery.

waria in the study had syphilis. It would be tricky because the current health system forced waria into men's wards. That didn't go down well with people who thought of themselves as women.

On a visit to Jakarta several years later, I dropped by for a gossip with Lenny. She counted off the latest test results – of eleven waria who had drifted into town from the outer islands of Sumatra and faraway Maluku in the previous few weeks, eight had tested positive for HIV. 'And three of my girls have died since last Sunday. Not a great week.' Lenny doesn't even bother to say what they died of. These days, waria don't seem to die of anything but AIDS.

These very high infection rates have important implications for clients, obviously. Not least because some of the men who buy sex from waria say they do it because they want to reduce their exposure to female sex workers, who they think have AIDS. In fact, a waria is between ten and twenty times more likely to be infected with HIV than a female sex worker in Jakarta, and the discrepancy is even greater in other cities.

One of the reasons that so many waria are infected with HIV is that they are often the receptive partner in anal sex, the 'bottom'. Of all sexual practices, receptive anal sex is the one most likely to lead to a new HIV infection, assuming that the insertive partner, the 'top', is himself infected. Waria get HIV from their clients – the men we were completely ignoring in the national HIV estimates. We had assumed that these men were always the 'tops'. But it turned out that nearly a third of waria had been paid to be top while the clients took the bottom role. In Nancy's view, this role swapping is another strike against sex change operations. Bad enough that they deprive you of orgasms. But they also deprive you of business, because without your equipment, you'll lose out on any client who wants to be bottom for a change. I didn't say this to Nancy, but I was pretty happy about that. One of the reasons that HIV spreads more quickly in anal sex between men than in any other type of sex is because men can switch roles. People who act as a bottom are highly likely to get infected in anal sex, regardless of their gender. If they have the equipment and the desire to act as top as well, they are highly likely to pass on that infection. Nancy may think sex change operations are bad for business, but they are actually good for slowing the HIV epidemic.

There was so much to try and understand just within the world of waria and their clients. But that was only one feature of the sexual landscape of Indonesia. It was just as important to try to understand how that little corner fed into the broader panorama. We learned from Fuad, the straight rent boy with the female sex worker girlfriend, that guys who buy sex from waria can be eclectic in their taste for partners. When we got around to counting, we found that over 60 per cent of men who said they bought sex from waria were married, and roughly the same proportion said they had casual girlfriends. Fully 80 per cent of clients of waria said they also bought sex from women, real women with two X chromosomes.

Sexploitation

HIV is very good at exploiting the diversity of human nature. It uses people like Fuad, who have lots of different risks, as a vehicle to get from groups with one type of risk to groups with some other risk. Let's suppose for a moment that Fuad is infected with HIV. Anal sex is a good way of passing on the virus, so it is quite likely that Fuad would infect some of the men who buy sex from him in the shadow of the Finance Ministry. The person he has sex with most frequently is his girlfriend, and she's also the one he's least likely to use a condom with. So if Fuad is infected, it is quite possible that his girlfriend is, too. And she's a prostitute who walks the streets; in the landscape of most of Asia's HIV epidemics, street-based sex workers report the lowest condom use. So if she gets infected, she may well pass the virus on.

Waria are unusual in the mosaic of prostitution in that they work independently. They have their cell mothers and their social organization, but they are not beholden to pimps or brothel owners. They can do their own negotiating because, for all the high heels and lip gloss they are, after all, men. Few clients would threaten a waria or cheat one: they are rightly wary of getting whopped over the head by a cute imitation Chanel purse, or having a red leather stiletto driven into their groin. We've tried to get waria to give assertiveness training to female prostitutes, but it hasn't worked very well. For women, the picture is just too different.

Female sex workers are less visible than waria to the casual traveller in downtown Jakarta, but there are far, far more of them. The city's health department believes there are around 28,000 women selling sex in the Indonesian capital, ten times as many as there are waria. At one end of the scale are the minor celebrities who charge thousands of dollars to provide discreet services to visiting dignitaries. At the other end are the ageing crones turning tricks at the back of the dockyard for a dollar or less. The market is pear-shaped – much heavier at the bottom end. But at every level, women tend to work within some kind of structure of 'protection', for which they pay a proportion of their income. If you want to flood the sex industry with HIV prevention services, it's important to understand how the local market works.

Like much of Asia, Indonesia has a schizophrenic relationship with its sex industry. There's no law against selling sex, though pimping is illegal. From the early 1960s, the government took a very pragmatic approach, concluding that what men wanted to buy, women would want to sell. Since the public frowned on 'women without morals', it seemed easiest to squash them together more or less out of sight in red light districts. Health officials were glad of this because it made it easy to round people up for health checks. It allowed the more conscientious to provide treatment for sexually transmitted infections, and the less conscientious to make money out of giving the girls injections of antibiotics, vitamins, sometimes even saline solution, in the name of 'STI prevention'. Distributors of condoms and stout beer, believed to increase virility, were glad because it reduced the scatter for their sales force and delivery vans. Clients were glad because they got 'one-stop shopping' – a huge choice of girls, music, cuisines all in a very small area.[4]

Then came democracy. Several Islamic political parties were carried forward on a rising tide of Muslim populism. More women started wearing *jilbab* headscarves. The new-found piety did not do much to change sexual behaviour though.* Looking back at my colleagues' early research it seemed that the proportion of men buying sex had remained steady since the mid-1990s, dipping only during the acute financial crisis of 1998. Most of these men are ordinary working guys with families. We asked almost 4,000 clients of sex workers in different parts of Indonesia if they were married; over half of them were.

* In 2002 there were two runaway bestsellers in Indonesia. One was a book of sermons by the popular Muslim preacher A. A. Gym, the other was Moammar Emka's *Jakarta Undercover* – true stories from an after-hours world where rich Indonesians eat sashimi off the bodies of naked prostitutes and gamble in all-nude casinos.

We see this pattern repeated in country after country around Asia: men tend to be slightly more likely to buy sex when they are single, but the majority of prostitutes' clients are married, simply because more adult men are married than unmarried.[5] Those men who can afford to will often buy sex even when they can have it for free at home.

The piety boom did not stop men buying sex, but it did provide Indonesia's populist politicians with a wonderful new Grand Gesture: running bulldozers through the edifice of sinful sex. All over the country, red light districts were crushed by the new demagoguery.

One of the largest and best-known red light districts was Kramat Tunggak in East Jakarta – a complex that provided work for some 3,000 prostitutes and a larger number of industry hangers-on. As it fell to the wrecking ball at the end of 1999, Jakarta Governor Sutiyoso declared that the closure was a signal from God. 'We are reminded that this is wrong and we should go back to the right path,' he thundered. To help the people of Jakarta find that path, the city built a vast mosque on the site where the dens of desire had stood.

What happened to the women whose livelihoods were wrecked by Sutiyoso? Some doubtless went back to their villages in West Java or further afield, taking any infections they might have with them. Some drifted on to the streets, where they continue to ply their trade without the health checks, clinics, or easy access to clients and condoms that Kramat Tunggak provided. Many, however, simply migrated a few kilometres north, setting up new clusters of bars and brothels in areas that were not in anyone's back yard.

Rawa Malang is one such place. To find it, turn right out of Jakarta's giant port complex and join the nose-to-tail queue of container trucks that lumber along the dusty, potholed road as far as the eye can see. At a small tin mosque, turn left and cling to the narrow dyke that threads its way between two huge, oily expanses of water perfumed with excrement, petrol and festering swamp life. As I passed by one afternoon, a ferryman punted two kids and their bicycles home from school. They live in a shack that squats on trembling stilts above this unctuous pool. I didn't envy them.

Just when you're thinking that no pleasure could grow in such a place, the swamp on the right solidifies into a little island of shacks and houses, piled on the site of an old cemetery. Turn off the road past the watchman sprawled on a bamboo bed, his belly naked to the humid breeze, and plunge down the embankment. You're in Rawa Malang. You can guess what goes on here by looking at the window display of the dusty medicine shop that sits at the entrance to the complex. 'Blue pills', a local knock-off of the impotence drug Viagra, jostle for space with traditional herbal 'strength medicine', said to have the same effect. There are bottles of the energizing tonic 'Extra Joss', marketed firmly at men needing a lift, and rows of antibiotics that are ineffectively prescribed for sexually transmitted infections. There's not a condom to be seen. Perhaps the store owner thinks it might reduce the market for the antibiotics.

The complex is divided into narrow alleyways, each lined on both sides with 'entertainment establishments'. Few are much larger than the sitting room of a run-of-the-mill suburban home. A large part of the available space is taken up with giant sound systems that pound out *dangdut*, Indonesia's music of choice, which mixes the why-did-he-leave-me-when-I-love-him-so lyrics of country and western with the insistent *dang dut dang dut* beat that gives the music its name – a sort of Dolly Parton/Technotronic mash-up. I went there looking for the head man of the area, Mr Lukman. In the oppressive heat of mid-afternoon, there was no one much around, just a handful of military types weaving to the *dangdut* in a mildly alcoholic stupor, watched by idle sex workers. Soldiers and policemen are such frequent clients of prostitutes throughout South-east Asia that we often include them as a high-risk group in our behavioural surveys.* It was a foregone conclusion that when the boys were done with their weaving they would take the girls into one of the little cubicles out back. The girls sat around picking their nails in boredom, waiting.

One mid-afternoon client told me it was good to get in early in the day, when the girls are still 'clean', and less likely to pass on disease. But most don't come until they've clocked off their work at the port and done their evening prayers at the mosque. Then the sound machines thud into service and the girls begin to compete for clients. By that time the school children who dominate the alleys in the afternoon will have gone to bed, and the itinerant salesmen flogging lipstick and plunging necklines will have been replaced by people touting flowers (for your beautiful lady friend, sir). The girls will no longer pick over their manicures. They will dance, and chat, and smile, and wheedle. They'll probably drink and smoke, and then they'll dance and wheedle some more. Eventually, they'll make a sale, and take their client into a back room. When they're done, they'll take 50,000 rupiah off him, around five dollars. They'll pocket three dollars and hand the other two over to the guy in the corner, the one with

* A note to this effect in a report on AIDS in Asia drew the comment, from a Harvard-based colleague: 'As this seems counter-intuitive would suggest a sentence explaining how this is the case.' These are young men stationed far from their families in a macho culture with rampant hormones, petty power, smart uniforms, ready cash and easy access to a bouncing sex trade. Counter-intuitive?

the sunny smile and the impressive biceps emblazoned with dragons and tigers.

Dragon-Tiger is neither the brothel owner (that's a Chinese businessman who comes on Saturday nights) nor a pimp, exactly. In Indonesian he's referred to as a '*preman*' which is often translated as 'thug'. But his role is far more delicate than thuggery. He takes care of the brothel owner's interests (hence the two-dollar cut, out of which he'll be paid for his own services). He provides security, makes sure that the girls are nice to clients, fields orders from regulars, and tries to ensure some kind of rotation among the one-off clients, so that no girl is left for too long without an income. Prostitution is like everything else in Indonesia, organized on a tight hierarchy that trickles patronage and petty power down the line, and in exchange sends money back up to a smallish number of kingpins at the top. If the system works well for patronage and petty power, I reasoned, couldn't we use it to distribute condoms and STI screening too?

FHI's sex trade team seemed to be stuck on a model of 'peer outreach' – getting sex workers to educate other sex workers. They'd been at it for six years or so, but condom use was still stubbornly low. As I began to unravel the mysteries of the sex trade, I thought we might do better by pulling the levers of power that already existed. It wasn't really my business to be wandering around Rawa Malang trying to figure out who pays whom for what, but I could always use the excuse of 'qualitative research' to stick my nose into my colleagues' business. I stopped in Dragon-Tiger's shop for a drink, and chatted with Nana, a thirty-four-year-old who worked to support four kids she rarely saw. They lived with her first husband's mother in East Java, but Nana wanted them to have a better education than she did, growing up in a poor ethnic Chinese family in Kalimantan, the

Indonesian part of Borneo. The man she now referred to as her husband was a sailor she met in Rawa Malang, but he appeared to be only an occasional visitor. Nana was not that interested in talking about her family; she'd rather chit-chat about the economy, the sex video of the politician and the dangdut starlet (filmed on said starlet's mobile phone) that was making the rounds, the breakdown in health services, anything other than herself. The impersonality had become ingrained over the years of chatting to clients and keeping a distance from them. It is when the distance breaks down, when a partner goes from being a client to being a 'boyfriend' or a 'husband', that things get blurry. Boyfriends tend to pay less, husbands not at all, even when they are essentially a prostitute's most regular client. If a woman is getting an emotional pay-off rather than a financial one, all well and good. More problematic for the one-track-mind health worker is that condom use tends to fall precipitously with 'special friends'. Men are three times more likely to use a condom with a woman they describe as a sex worker than someone they say is a girlfriend, even when the girlfriend is actually a favoured hooker.[6]

The pattern is more or less universal, and it is eminently understandable. It was explained to me succinctly by a former sex worker in Chicago, in the United States – about as far as you can get from Rawa Malang. 'Well of course I use condoms with all my johns. That's work,' said Leticia Brown. 'But with my pimp, that's different. He's my man, that's love. I ain't going to use no rubber with him.' Leticia was infected with HIV by her pimp. 'And then I buried him.' She now works as an HIV prevention counsellor, but she is not optimistic about persuading sex workers to use condoms with their intimate partners. 'Everyone's got to make a difference between work and home life.'

Paying the price

For many men, buying sex *is* part of their work life. Not the men we capture in our HIV surveillance system. For them buying sex is pure pleasure – a well-deserved treat after a tedious fortnight cooped up on a cargo ship on the high seas with a bunch of sweaty sailors, or a good way to stretch the limbs after wrestling an oversized truck over scrappy, undersized roads for eighteen hours. But there's another type of client, the ones we rarely capture in our surveys, who buy sex on expense accounts. These are the flashy traders and businessmen with clients to impress who crowd the slick, upmarket nightclubs of Asia. I first met them when I walked wide-eyed into Hong Kong's glittering nightclubs as a rookie reporter for Reuters. I was covering business at the time, and was curious about the planned stock-market flotation of Club Volvo, the mother of all pick-up joints. I've been in a lot of pleasure palaces since then, but none has matched the kitsch of Volvo, a club that makes you feel like you are trapped inside a wedding cake iced in brass, mirrors and cheap satin.

In the mid-1980s, this sort of place was a honeypot for men in plastic shoes and greasy comb-overs, the tell-tale signs of a Communist Party apparatchik from across the border in mainland China. These mandarins were the gatekeepers to China's increasingly solvent markets; they were the guests of Hong Kong businessmen who hoped their hospitality would be rewarded with contracts. The bad comb-overs were ferried across the dance floor in a pink and gold vintage car – that's how big the club was – to tables where their hopeful hosts waited with a small selection of the club's thousand or more temptresses.

On each table sat a computerized time-puncher. When a lithe demigoddess attached herself to a table, she punched in her number and the clock started running. Clients paid the club US$45 an hour for her flirtations, and would tip her too, of course. If a client wanted to take a girl out, he had to pay the club for her time until closing – 3 a.m. What he paid the girl after that was their own business – these clubs did not condone prostitution, obviously; they would never discuss prices for anything other than the charming company of a young girl in a satin gown slit to the waist.*

I wondered what the embryonic 'ethical funds' would have to say about buying shares in Club Volvo – after all, the club was positively discriminating in favour of women and youth in its workforce. I never found out. A week after the flotation was announced by Club Volvo investor Ronald Li, who happened to chair the Hong Kong stock exchange at the time, the world's stock markets went into a swan dive. On Black Monday, 19 October 1987, Hong Kong's investors saw a third of their paper wealth go up in flames. Since Volvo's business grew like a lotus from the slush of brokers' expense accounts, this was not good for business. 'It's a matter of weighing up the cost of taking clients out to clubs against the amount of business they are likely to give you,' a chastened US broker lamented to me at the time. 'Even soft bribery has to be justifiable.'[7]

The glitzy nightclubs blossomed again along with Asia's boom economies, and by the time Reuters transferred me to Jakarta at the end of 1988, several of Jakarta's finest clubs were booming. I used to go to them sometimes in those days, while schmoozing some tycoon for information about his business. But ironically when I came back as an observer of the sex industry a decade

* Regulars got the timing of their arrival down to a fine art. 'Too early and it makes for a very expensive evening. Too late and you get left with all the dogs,' one of them explained to me at the time.

later, I barely set foot in the posh places. From an HIV point of view, they didn't seem that important. Condom use rises with price. So does the ability to pay for decent health care. We had more than enough work to do in places like Rawa Malang. Girls who were pocketing over US$100 a night could take care of themselves.

Getting down

As I learned more about the sex trade, I learned that the prostitution pear really only applies to the heterosexual trade. In male prostitution, there's virtually no high end, at least in the countries where I worked. These days there are a few 'spas' opening up here and there, places with cappuccinos and glossy magazines, the porn on flat-screen TVs. But on the whole, even my poshest gay friends buy sex from the same massage parlours and sleazy discos and cruising parks as everyone else.[8]

In Indonesia, one of the richest gay guys I know keeps a series of Excel spreadsheets ('ForFriends.xls') that list gay brothels, massage parlours and cruising areas in various cities in Indonesia. He doles these out liberally to wealthy associates who share his interests – a great 'key informant' in our qualitative research. In the course of that research, my colleague Philippe Girault and I went to a very large number of the 100 or so sites listed on the Jakarta Friends spreadsheet, which includes helpful details such as 'rooms are a bit hot', 'next to the police station' and 'chinese moslem man, very big size dick'. Although many of the boys on offer were plenty cute, the places they worked were almost uniformly downmarket. Philippe is a relentless Lothario who looks like Gérard Depardieu cast in the role of an ageing rugby player and who cheerfully admits to liking his men a bit rough. But even he was shocked by some of the most eagerly patronized places on the Friends list. A particular favourite was a massage parlour where the rent boys are blind. He messaged me in the middle of our mapping exercise: 'This is a very "glauque" and incredible scene. Never seen that before.' *Glauque*, a wonderful French word with no proper English equivalent; it encompasses sleazy, seedy, sordid but somehow bordering on the excitingly wicked – the feeling you get after getting sucked into a vortex of passages barely wider than your rugby-playing shoulders, feinting past piles of decomposing garbage and being disgorged, finally, in front of a tiny shack with a tin roof where blind boys sit 'watching' a blaring TV, waiting to be steered with a customer into a cubicle partitioned with cardboard and worn polyester curtains to feel their way to business.

At about the same level are the boys who hang around the inappropriately named Grand cinema in Jakarta. A cavernous place where cockroaches outnumber patrons by a zillion to one, it distinguishes itself by showing something close to salacious movies. But its principal asset is that it is dark. Here's how to take advantage of that fact. Buy a ticket. Go in to the once-grand marbled hall. Raise your eyebrows at one of the teenaged boys hanging around the lobby area (they stay on the inside so that they don't have to buy a new ticket for every new customer). Pick one of the subdivided screening rooms at random. Plunge into the darkness. Sit down, making sure there is an empty seat on either side. The boy following your eyebrows will sit down next to you and plunge, either with his hand (for 10,000 rupiah, about a dollar) or with his mouth (up to 50,000, if you're feeling generous). You can leave whenever you are ready. Not many people watch a movie all the way through at the Grand.

The Grand is conveniently situated close to a major bus station which brings working-class men into the overcrowded capital from the rest of Java. There are almost 1,000 people squeezed into every square kilometre of the island of Java, compared with around 240 in the UK and a paltry 31 in the US. Most live in what-will-the-neighbours-say proximity – in some villages women's contraceptive choices are still posted on the health centre walls, a relic of the 'contraceptive safari' days of family planning by intimidation. Everyone knows everyone: the potential for anonymous anything is slim, and the potential for anonymous sex close to nil. The ask-no-questions darkness and rock-bottom prices of the Grand are precious commodities for men whose village norms do not overtly embrace sex between men. They can tip off the bus, indulge their needs and desires at an affordable price, and then get on with business in the big city without much fuss.

Better still, in some boys' books, is to be on the other side of the equation. 'Sometimes I have to slap myself to make sure I'm not in heaven yet,' said Rudi, a pale and beautiful ethnic Chinese boy who works at the Parahiangan massage parlour in central Jakarta. Rudi grew up in a small coastal town in North Sumatra. He told me he had always been attracted to men, but didn't know sex with a man was possible. 'I was so dumb, I didn't even know I was gay.' Except that he didn't say gay, he said '*sakit*', which means sick, ill, unwell, and which has become slang for gay. In slang, the opposite of 'sakit' is 'normal'. Rudi came to Jakarta in his late teens, and saw ads in the back of the lurid tabloid *Pos Kota* that seemed to suggest that there were men available for sex with men. 'Well, anything can happen in the city, I thought, so I called one of the hand phone numbers and next thing I was having sex with a guy and mmmm, it was nice.' It didn't take Rudi long to figure out that if money was going to exchange hands for sex, he'd rather be on the receiving end. 'And now I get paid to do something I've wanted to do my whole life without even knowing it.' Rudi grinned in disbelief at his good fortune.

Then, to my astonishment, he asked for advice on what he should do to become 'normal'. 'One day I want to go back to my village and get married and have children. It's too much of a pity for my parents if they don't have grandchildren. Tell me, how long do people usually stay "sick"?'

Whatever his long-term fantasies, there was no doubt about Rudi's sexuality. He loves having sex with men, and it is an added bonus to get paid for it. But many men who sell sex to men do it for the same reason that women sell sex to men: for the cash. Like Fuad, who works close to the Finance Ministry a few blocks north of the Grand cinema, they have no trouble differentiating between what they do for income and what they do for pleasure. Half of the male sex workers we spoke to had sex with women by choice, never having sex with men unless they were paid to do so. One in seven of the rent boys (selected because they sold sex to men) said they also sold sex to women.[9] That surprised me. Who were their clients? In four years of asking around, I never met an Indonesian woman who admitted to buying sex. The boys said they were paid by married women neglected by philandering husbands, but I was never able to get any real idea of who these women were. The rent boys reported nearly 60 per cent condom use with female clients, which left me concerned for the other 40 per cent. I wondered if they knew that their fancy-boys were also selling anal sex to men.

Love blossoms

There are doubtless people with genuinely eclectic sexual tastes, but the more we learned about the country's sexual landscapes, the more it seemed that there were two faces to bisexuality in Indonesia. On the one hand there are people like Fuad, men who would rather have sex with women but who build up their financial capital by selling sex to men. On the other hand there are the people who buy sex from Fuad, men who would rather have sex with other men but who build up their social capital by having apparently 'normal' sexual relations with women.

The second group appears to be diminishing over time, as it becomes more acceptable, and more possible, to be openly gay. When I worked in Indonesia as a foreign correspondent in the late 1980s and early 1990s, there was no gay scene in Jakarta. A few young Indonesian guys lived with their Western boyfriends, but I didn't know any openly gay Indonesian couples. Nice boys got married. A few flamboyant individuals were assumed to be 'not the marrying kind' but you could count them on one hand: a designer, a senior tourism official, a dancer. There were no gay bars, though the Indo-Western couples would occasionally dance side by side at the Tanamur, a heaving bordello-cum-disco that admitted all behaviours. They went largely unnoticed, drowned as they were in a sea of heterosexual lust.

I was a bit of a regular at the Tanamur in those days. And the object of my heterosexual lust was a smooth and honeyed Javanese man, Bhimanto Suwastoyo. Bhim was my counterpart at the rival news agency Agence France Presse, and a wonderful dancer. He was unusually demonstrative in public, especially for a well-bred Javanese Muslim. We danced indecently closely, we held hands, we found excuses to go and cover the same stories in far-flung islands. But after a week, a month, a whole rainy season of hot nights together, the public hand-holding never progressed to anything more steamy.

I asked Bhim if he was gay. He looked horrified. Of course

not, he said. We went on holding hands, and I went on being crushingly in unrequited love until I left Indonesia, two long years later.

When I came back to Jakarta for a quick visit in 2000, Bhimanto picked me up at the airport. We drove through a city I barely recognized, a city of glass skyscrapers and new toll roads and billboards advertising the myriad political parties of the post-Suharto era. The Tanamur was a morgue. Bhim took me to his new favourite joint, an Italian bar called Prego. As we walked up the stairs, he said casually, 'By the way, did I mention that it's a gay bar?' And then we were in a room packed with sculpted male bodies. The music thudded, the crowds embraced, the bartender had put a Chivas in Bhim's hand and someone else's hand was on his backside. It belonged to Bhim's partner Stephen.

Bhim's isn't just a run-of-the-mill coming-out story. He can now admit to being gay because a whole city is coming out, a whole continent in fact. The scene is being repeated in major cities all over Asia.* Men who just a few years ago would rather die (or marry) than admit to being gay are gathering in bars, dancing cheek to cheek and going home with new boyfriends. It's not just in capital cities, either. In the frozen north-east of China, in the bleak city of Harbin, colleagues describe a similar coming out. In 2002 just 58 per cent of men who said they sometimes had sex with other men identified themselves as gay. Four years later, 80 per cent were out and proud. Lots more were shacking up with other men, too – up from 12 per cent to

* The *Far Eastern Economic Review*, for decades Asia's leading news weekly, carried a cover story on Gay Asia in its final issue in October 2004: 'Coming out, cashing in'. The issue included a side story about Bhimanto, Stephen and their adopted son Arya, as an illustration of how social mores in Indonesia had changed.

41 per cent.[10] These changes are at once heartening and worrisome. Heartening because no one should be condemned to live a subterranean sex life. Worrisome because an active gay scene creates instant opportunities for men to meet new lovers. As we saw during the gay liberation of the 1970s in the US and Britain, an active gay scene concentrates risk. It gives more people the opportunity to have sex with more different partners, and it gives HIV a chance to spread.

When gay bars and bathhouses were springing up in New York and London and Sydney in the 1970s, we didn't know about HIV. Now, as they blossom across Asia, we do. We know, too, that you can have a very active gay scene and control HIV transmission as long as everyone wears their latex. But at the start of a country's gay liberation, when gay men are just relieved to be able to burst out of the closet, no one seems to think too hard about things like condoms. It's about getting out there, expressing yourself and having fun. 'Community activism' seems a long way down the road.

The high life

Sex is something I know about, something I do, something everyone I know does. But there's more to HIV than just sex. HIV also goes hand in hand with drugs, and drug injection was not really part of my universe. If it hadn't been for a cute boy in the WHO canteen back in Geneva, drug injection would barely have been on my radar screen. When I arrived in Jakarta, I had never even spoken to an injector.

If we thought about drugs at all, it was usually in the context of risky sex. And the two do go together. All drugs (except perhaps tobacco, but definitely including alcohol) make life harder for those who work in HIV prevention, because all drugs make you temporarily stupid. Many drugs make you horny, or lazy, or careless, or all of the above. And all of those things make you more likely to have sex, or less likely to do sensible things like use condoms when you do, or both. Even in overwhelmingly Muslim countries such as Indonesia where alcohol is not deeply embedded in the culture, we see a strong correlation between drinking and risky sex. It holds for men and for women, for older and for younger, for buyers of sex and for sellers. The correlation between other illegal drugs (such as heroin and cocaine) and stupid sex is stronger still.[11]

There is no shortage of recreational drugs in Asia. The continent is awash with them. Ecstasy ('X') is big, as are methamphetamines, known in Indonesia by the name of a Japanese hot-pot dish, *shabu shabu*. If you want to slow down rather than speed up, there's a fashion for 'Special K'. Known to vets as ketamine, it leads a double life as a horse tranquillizer. And then there is the grand-daddy of all Asian drugs, heroin (or 'smack', to its friends). All making people do stupid things.

One of the stupidest: using a filthy needle to inject a lethal virus straight into the bloodstream.

In Indonesia as in China, Vietnam, Thailand, Burma, northern India, Pakistan, Nepal – oh, and Argentina, Chile, Russia, Ukraine, and let's not forget Spain, Portugal, Iran and even parts of the urban United States, Scotland, the list goes on . . . drug injection is *the* major engine pushing HIV into new bodies. In a few countries drug injection was what was left after they had dealt effectively with other risks such as a lively commercial sex industry (Thailand) or a pre-condom gay scene (New York City). But in many, many countries drug injection is the sine qua non of the HIV epidemic. And new communities of injectors are coming on stream all the time.

Heroin use varies from country to country. It took root decades ago around the fields of poppies that provide the ingredients for heroin. In the north of Thailand and India, in Afghanistan, Burma and south-western China, there are whole generations that grew up with drug injection all around. In other places, sticking needles full of intoxicants into one's body for the sheer pleasure of it is a very new phenomenon.

When I was a journalist in Jakarta in the late 1980s, a handful of middle-aged Indonesians who had once played in jazz bands in Amsterdam used to shoot up. By the time I came back a decade later, I could walk out of my house, clap my hands, and score heroin for around US$3 a hit, faster than the more legitimate street vendors could whip me up a plate of fried rice.

Does that give the impression that I was living in some edgy area so that I could commune with the people I was hoping, in some distant way, to help? Not a bit of it. I was one block from the governor of Jakarta's house, two blocks from the US ambassador, back-to-back with the former vice-president of Indonesia, and just four doors down from the current vice-president. In front of our houses paraded a steady stream of Pajeros, Jeeps, BMWs and Mercedes, often driven by kids in the uniforms of Jakarta's posher private high schools, always pounding out window-rattlingly loud techno music.* They came for the fried rice, the best in Jakarta, some would say. But they came also for the panoply of drugs on offer, cannabis or shabu shabu, ecstasy or smack. Increasingly, smack.

Why is heroin the drug of choice for so many in Asia? For a

* This was too much for my neighbour the vice-president. He fled to a quieter neighbourhood, leaving his cousin to put up banners around his abandoned residence come election time.

start, it is cheap. In the official price tallies kept by the United Nations Office on Drugs and Crime, Indonesians paid a retail price of just US$34 a gram in 2004 for heroin.*

But smack users also like the fact that you get that nice, warm, flooded-with-light-and-calm feeling without too much disruption, without feeling all hyped up, or all dulled down. I learned that while chatting to Desi, an addict in her late twenties. 'If you take shabu you can't sleep, and if you smoke dope you can't work,' Desi told me. 'With smack, you can do pretty much anything.' She took a long pull on her cigarette. 'At least at first.'

Desi is a long way past 'at first'. Her skin is still radiant, her body lithe. But Desi's been using drugs for a decade and heroin for seven years, the last three of them injecting. Like many other injectors she started shooting up as a way of economizing. Smoking heroin – 'chasing the dragon' – is nice enough, but you inevitably send part of your precious high up in smoke. By injecting, her friends told her, she could get more high for her money because it would all go into her veins. 'And it is sort of true,' Desi says. 'But in the end it doesn't save you any money. You always buy all the drugs you can, until your money runs out. So in the end you just get high more.'

That 'more drugs equals more high' attitude can be catastrophic among new injectors. And in Indonesia, where heroin

* Heroin in Indonesia is of variable purity. In Hong Kong it's pretty high, 62 per cent pure heroin on average, and still really cheap at US$46 a gram. In the producer nations heroin is cheaper than printer ink – US$7 a gram in Burma, for example. Only micro-managed Singapore bucks the norm. There, heroin users have been paying upwards of US$130 a gram for a drug that is just 5 per cent pure. That means they pay nearly forty times as much for their high as their Indonesian neighbours, and nearly ten times as much as users in the United States or the UK, where prices are lower, at around US$95 a gram, and purity is much higher. Source: *United Nations Office on Drugs and Crime, 2006.*

has been easily available for about as many years as democracy, virtually everyone is a new injector. 'These kids have no idea what they're doing,' said David Gordon, a 'former criminal type' (his words) from Los Angeles who now runs a treatment centre for drug addicts in the hills east of Jakarta. David has untold jail time and fifteen years of living as an addict under his belt, so he knows his stuff. 'They really like getting high. Who doesn't? But they think that if they take more drugs, they'll just get higher.'

What they get, very often, is dead.

An over-enthusiasm for injecting combined with very variable drug purity means that new injectors often misjudge the amount of toxins their body can cope with. The result is very high rates of overdose.

I have to say that I struggled a bit to get my head around this heroin business. Like most people who went to high school in the 1970s I knew how to roll a joint. Like most people who hung around Studio 54 and the other cocaine-fuelled clubs of New York City in the late 1970s, I knew how to roll up a hundred dollar bill. But I didn't know anyone who stuck needles into themselves for fun. As I began to turn up the curtain on the world of drug injectors, I'd talk about it to my nice, middle class Indonesian friends. To them, there was nothing extraordinary about my junkie stories. My friend Rosa, a highly paid executive with Big Pharma, lost her brother to smack. She had the sympathy of Nungky, an architect who once swung up to my house in a Mercedes. Nungky's a good architect, and her firm is doing well, but she was under thirty at the time. A Mercedes? It turns out that it was on loan from her mother, who feared that if she left it around the house Nungky's heroin addict brother would use it for assignations with his dealer. Another friend, Sari, had a happier story – after many, many years shuttling in and out of rehab her brother got clean ten years ago. 'But his addiction completely traumatized the family. I remember thinking: everyone else is overdosing, why can't he? I just wished he would die.'

I commented on the 'junkie in every closet' phenomenon to a neighbour I'll call Sandra.[12] Sandra is a determined woman who organized a neighbourhood watch committee that eventually tossed the drug dealers out of our street, though only after ten o'clock each night. She launched straight into the story of her own cousin, who had just been arrested in front of our local mosque for dealing drugs, husband in tow. 'They had their two daughters in the car so that they could look like a nice, middle class family,' said Sandra in some disgust. But these days, being a nice, middle class family is in no way incompatible with being involved with drugs. Drug use is such a norm now that in our surveys of Indonesian high schools more kids report using illegal drugs than having sex.[13]

Sandra's cousins were locked up in police holding cells. The eldest daughter, a middle school student, was understandably upset and solicitous for her parents' mental welfare. She sent in some reading material with an uncle who went to visit. At a routine door check as the uncle went into the cells, it was discovered that this thirteen-year-old had thoughtfully put a care-package of heroin inside the book for her mother.

To learn more about middle class drug addicts, I went to the rehab centre run by ex-junkie David and his wife Joyce. In a villa that was built to allow the gilded families of Jakarta to escape the heat of the city, the gilded families of Jakarta now dump their most troublesome members. Some of them will follow the detox and recovery programme for several months and go home clean. Most will be back before too long.* Or they'll get dumped even further away, in one of the just-say-no boot camps run by militant Muslim groups which are mushrooming across Java as the need for drug treatment grows.

There was a certain sameness about the boys at the centre. Long hair, tattooed torsos and low-slung shorts are the plumage of the species. Occasionally, an exotic migrant flies in from elsewhere. One youngish man with a mop of curly hair and honey-coloured eyes told me he had come from the Maldives for training. Being trained to do what, I asked? 'Live without drugs,' he replied, not overflowing with enthusiasm for the idea.

He echoed Desi's preference for heroin, because you can lead a more or less normal life, hold down a job, all of that. So what was his job?

'Airline pilot.'

He flew jets for Malaysia Airlines, among others. On which, as you are coming in to land, the stewardesses stand in their ugly turquoise and lavender batiks, reminding travellers of the severe penalties for drug use in Malaysia.

Of course drug injection in Indonesia is by no means confined to nice middle class kids who go to comfortable rehab programmes as a career break from their jobs flying planes. Now that smack is so cheap, it has cascaded down the social totem pole, wending its way behind the lace curtains which working class communities use to block intrusions from the rest of the slum.

* Voluntary detox centres tend to estimate relapse rates at between 70 and 80 per cent. Where police arrest junkies and send them to compulsory detoxification centres, as in China and Vietnam, relapse rates are typically much higher. Public security officials in China's drug-laden Yunnan province report relapse rates of between 90 and 96 per cent. My Chinese colleagues could not find a single person currently injecting drugs who had not been through rehab at least once; most had been 'treated' between three and seven times. Source: China UK AIDS Prevention Project, 2003.

Here, as generations pile on top of one another in one-room houses, finding room for country cousins as best they can, injectors can't shoot up in the quiet of their own bedrooms. They have had to find other places in which to get high. There are the sinister, urine-stained spaces beneath thundering highway flyovers, but a current favourite in Jakarta is the railway station. The tracks trail through many of the poorer areas where drug use has become a feel-good alternative to the boredom of unemployment. You won't get picked up for loitering on a station platform; it's de rigueur, given the erratic nature of the train services. There's cover against the worst of the sun and any unexpected tropical dousings. And there's almost always a public bathroom that affords at least a modicum of privacy for the actual injecting. Overall, a pretty good place for junkies to hang out.

I went with long-haired Made to Tanah Abang station to meet some of Jakarta's less gilded injectors. There, squatting on a railway platform surrounded by the alleyways and canals of the Kampung Bali slums, I learned that stations have another advantage. A train pulled up and disgorged its contents: the bloke flogging green worms, that much-loved dessert made of gelatine and what? – petroleum by-products? – the harried civil servant sweating into her beige uniform as she raced to get home to feed the kids, and then the boy in the junkie uniform – long hair, tattoos, low-slung trousers, an inevitable cigarette hanging from his lower lip. After a short exchange of ideas with one of my companions, he disappeared into the public toilet. He emerged beatific a few minutes later, in time to climb back on the train and move away from bad company and evidence. The needle he used was back in place on the concrete ledge high above the washbasin in the station toilet. It was out of sight to anyone who was there just to urinate but easily available to any passing injector who might be past caring about the needle's promiscuous past. The only sign that our visitor had just shot up would be in his pin-sized pupils, and they are especially hard to spot if you've got dark Indonesian eyes.

Cheap though they are, drugs have to be paid for. Injectors with no jobs and no middle class parents to steal from had to come up with the cash one way or another. When we started counting, we found that one male injector in ten sold sex to pay for drugs in some cities. But if Toilet Junkie did that, should I count him as an injector who happened to sell sex, or a rent boy who happened to shoot up drugs?

Bit by bit, I began to understand that we'd have to work towards a surveillance system that took people out of the neat boxes we had wanted to squish people into in Geneva: 'male prostitute', 'drug injector', 'client of female sex worker'. A system that at the very least redrew the boxes as overlapping circles, a great Venn diagram of sex and drugs, desires and needs, hormones and money.

By the middle of 2002, we were moving beyond the qualitative research and starting to count things in earnest. In Indonesia, and indeed in most of Asia, it would turn out that everything overlapped with everything else.

3

The Honesty Box

All those years writing manuals telling people how to collect information on HIV and the behaviours that spread it, and I somehow omitted the most important point: anything that can go wrong will go wrong. As we wrapped up our qualitative research, rewrote questionnaires, and started fanning out to ask thousands of people about their sex lives and their addictions, I realized that it was one thing to chat with a couple of waria about their husbands, but quite another to ask a couple of hundred waria to let you take blood from them. Good HIV surveillance involves prostitutes and drug addicts, pimps and cops, blood and swabs from every orifice.* There is a lot that can go wrong.

Plan for errors, record them, fix them. If you can't fix them, own up. Because mishaps in research affect results, and wayward results can assume a life of their own, worming their way into speeches, newspaper reports and the public consciousness

* We often take urethral, vaginal and anal swabs to test for STIs other than HIV.

and leading to bad policy decisions. One way or another, they will come back to haunt you.

A few years ago, the *British Medical Journal* floated the idea of including an 'honesty box' in each of its articles – a space for researchers to record the 'warts' in their data, the little things that go wrong in the field.[1] Because let's face it, things go wrong even when you're going door to door asking people what they ate for breakfast. In HIV research you're going from brothel to gay bar asking about behaviours that are often illegal or embarrassing and collecting specimens to test for an unspeakable fatal disease in countries with erratic vigilante movements and an irregular power supply. The honesty box can fill up quite quickly.

To get good information about anything, you have to talk to the right people. You have to ask the right questions and record the answers correctly. If the interviewer is so distracted by the sex worker's cleavage that he ticks the wrong box, the data will be worthless. If you mix up the samples by mistake, sticking the study number for the short fat guy in the pink T-shirt on the tube of blood from the cute chick with the nose stud, you'll get nonsense. Later, when you've linked the information they gave you about their sex lives to the results of the blood tests, you could find you have virgins with sexually transmitted diseases or pregnant men. In the chaos of a nightclub or back alley, it's remarkable how often these things can happen.

People think of research as something that involves questionnaires and test-tubes, computers and laboratories. But most epidemiologists will tell you that the hardest part of any study is recruiting the right people. Indonesia's surveillance system already covered sex workers, and there was some behavioural surveillance among men likely to be their clients. Drug injectors seemed to be the engine driving HIV infection in Indonesia but we only had testing at one drug hospital in Jakarta, and there was hardly any information about behaviour. Waria, rent boys and gay men didn't exist as far as the national surveillance system was concerned. We had to find ways to reach all of these people so that we could collect specimens and information from them. But more importantly, we had to have some idea whether the people we collected information from were more or less the same as others in their group. Did the women tested in massage parlours all sell sex, or did they include those who just gave massages, with no extra services? If we couldn't answer such questions, we'd have no idea what our results meant. What, for example, was I to make of the category 'Etc.', which some provinces reported along with sex workers and pregnant women as a surveillance group? Who does 'Etc.' represent?

There are all sorts of guidelines to help figure out how to reach people at risk for HIV, most called something like 'HIV surveillance in hard-to-reach populations'.[2] In fact, with the exception of drug injectors, most of the populations that we are interested in when planning HIV prevention programmes are not that hard to reach. Take sex workers, for example. To make a living, a prostitute needs clients – a hard-to-reach prostitute would starve. And if her clients can find her, so can we. It just requires a bit of local knowledge and some mapping skills.

Here's a summary of what the manuals tell you to do. From people who know the market – people who sell sex and the things that go with it, that uplifting stout beer, condoms, virility potions – find out where the action is. Then map those places, visit them to estimate the number of people working at each site, and cluster them together if need be. At random, draw a sample of clusters. Now visit the ones that were selected, and pick people, again at random, to participate in the survey. You might pray that you don't draw Nana's stomping ground so that you don't have to tramp yet again through the swamps of North Jakarta. But you don't get to choose. Random is random.

In theory, Nana has the same chance of being included in the sample as Lili or Rita or any of the 28,000 other girls who sell sex in Jakarta. That means it is fair to assume that whoever does get included in the survey is reasonably representative of all 28,000. So you can assume the sex workers you have collected information from are the same as sex workers you haven't collected information from. If you ask a statistician.

Or not the same, if you ask a sex worker. Our tidy reliance on the manual was shaken apart by Ines. Ines is an intimidating beauty. She is tall and elegant, with an aquiline, almost haughty, nose and wide, liquid eyes. Her hair falls in glossy waves; she must have spent hours perfecting that casual flick that sends it rippling over her long neck and down her back. The flick manages to say 'come hither' and 'don't mess with me' in the one, smooth movement.*

Ines has a penis. One of Jakarta's better-known waria, she sells sex only occasionally, in a rather desultory sort of way. The first time I met her, she was wafting around outside the Italian Cultural Centre in Jakarta, a favourite cruising spot for the better-heeled sensation-seekers of the city. The bunch of daffodils she was cradling apparently came from a client. 'So romantic, don't you think?' But when she saw I was having trouble with my motorbike the daffodils got ditched and she was right in there changing spark plugs. I fretted about her manicure, but Ines didn't care. 'You know how much I got paid tonight? I'll have my nails done again tomorrow.'

* Meet Ines in a video on http//www.wisdomofwhores.com.

Just as well she's making good money – she has expensive taste. She came over to my house one night to meet Shanty Harmayn, a (similarly beautiful) film producer who was looking to cast a waria in her new movie. Ines was dressed with classical elegance – a modest black dress with a beautiful batik scarf and a discreet gold pendant. Her make-up was understated. I can't say the same of her consumption of my best single malt whisky. 'Mmmm. I usually only drink Chivas, but this is nice,' she pronounced, draining the last of the Laphroaig.

The conversation drifted to my work. We'd just finished our first survey of HIV, syphilis and risky sex among waria in Jakarta, the one that made me cry. The behavioural results had shocked me too, though in a different way. I had the impression from the qualitative research as well as from Nancy, Lenny and co. that waria were turning dozens of tricks a week, but the study showed they averaged only three. And since that figure came from 250 waria selected at random as the manual requires, it was certainly more accurate than the qualitative research or my vague impression based on chats with four or five friends.

'Three a week? You're insane!' snorted Ines. 'But then look at who you ask.' Ines left school at fourteen but she is not stupid, so I explained how, through our probability sampling, we are asking a representative sample of Jakarta's waria. Ines dismissed my lecture with a wave of her manicure and a flick of her locks. 'So clever, but so stupid,' she sighed. She explained, as if to an enthusiastic but slightly dim child, that a waria who is hanging around on a street corner to be interviewed by a research team is a waria who is not with a client. 'You're talking to all the dogs, *obviously*.' Not something that I learned in the lecture halls of London, not something that went into the surveillance cookbooks, but of course Ines is quite right. Our sample is biased towards the 'dogs', who get picked up less than the cuter girls. So the study results underestimate the true number of clients per seller.*

Ines's comments on that long, liquid evening prodded us into changing the sampling strategy. Now, in areas with a closely knit waria community as well as in some establishments with male and female sex workers, we work with the powers-that-be (the mami, the pimps, the brothel owners) to arrange off-hours times for data collection. The principle upside of this is that you are not cutting into people's work time, so there is less chance of talking only to the remnant sex workers who can't get a client.

Off-hours interviewing is easier on research teams, too. We were working to shift the behavioural surveillance work from universities to government agencies, and civil servants do not exactly line up to work from 11 p.m. to 3 a.m. in noisy, over-crowded, smoky rat-holes in the sleaziest parts of town. Forward planning can also minimize some of the vagaries of politics and the weather. These can make an established cruising area – and its established cruisers – evaporate overnight.

Or faster. Once during the waria study, the day my husband David reappeared from a long stint in war-ravaged Afghanistan, I had the audacity to take a night off. Bored by sitting around editing copy in comfortable Singapore, David had turned himself into an occasional 'fireman', dropping in on any conflict that happened to be raging. I was used to it and didn't worry too much, but I still thought his safe return from a war zone merited a celebratory night in. At around midnight, my phone rang. David rolled his eyes. I ignored the ringing; let Philippe deal with it. The Depardieu look-alike was on duty as

* Ines didn't get the part in Shanty's film, though she found much else to busy herself with. When I ran into her in 2007, she had passed qualifying exams and signed up for a part-time degree programme in human rights law.

study supervisor that night, and he was a lot more experienced than me at this sort of field work. Two minutes later, my phone rang again. It *was* Philippe. 'You have to come. There are cops everywhere. They think it's us. We're by the canal.' Who thinks what is us? But Philippe was too agitated to make sense. 'Just come!' And he hung up. David looked resigned as I dragged myself out of bed, picked up my crash helmet and headed out the door.

The research team had got stuck in a raid by the *trantib*, a sort of 'discipline police' that tries to keep the country morally pure. They were loading waria into the backs of vans to take them to a rehab centre, where they'd spend a day or two before hitting the streets again. The *trantib* were meeting some resistance, not least from Lenny and two other waria who were out there that night as interviewers. 'Do I look like I'm selling sex?' asked Lenny from under her *jilbab*. To add to the chaos, the *mami* of that area, Myrna – a big rival of Lenny's – was accusing the research team of bringing in the cops on purpose. Philippe, raging away in English, was not adding clarity to the situation. I pulled out some letters from the ministry of health, bedecked with official-looking stamps, and did lots of bowing and scraping. Eventually everyone calmed down and I could go back to my abandoned husband, who was by now fast asleep.

We resampled the raided area again a week or so later, when everyone was back at work. Sampling purists would probably sniff at this, because it deviates from the truly random. But sampling purists do not work on the streets of Jakarta at midnight.

Since then, we've learned to inspect the calendar for upcoming local elections when planning field work. It seems that nothing delivers votes quite like a law-and-order crackdown; sweeping the streets clean of sex workers and shutting down massage parlours is a great way for incumbents to increase their ratings ahead of an election. Large international meetings or sporting events are crackdown magnets, too. In Muslim countries we obviously don't do any data collection during Ramadan, when people abstain from food and drink from dawn until dusk, and abstain from sex from one end of the month to the other. Many sex workers go back to their home villages for family visits during this time.

These things are predictable. Others, such as tabloid headlines trumpeting the murder of prostitutes by abandoned pimps, or the random 'sweeping' of sinful districts by zealots from the radical Islamic Defenders Front, the FPI, can also wipe a carefully mapped area of its expected yield of survey respondents. There's less we can do about that.*

Predictable is what we like in epidemiology. Point number one for the honesty box must be that we don't always know how the people we are talking to represent any bigger population, but we are getting much better at controlling the unknowns.

Point number two is that we don't always know if the people we do talk to tell us the truth.

Yesterday's truths

There is an assumption that people automatically lie about their sex lives, and are even less likely to tell the truth about drugs.

* The FPI threat seems to have become more predictable of late. Eris, a friend who runs a nightclub, told me recently that the *preman* thugs who have always run the protection racket in her area have taken off their leather jackets, put on skull caps and grown beards. Because of the political sensitivities surrounding Islamic fundamentalism, adopting the uniform of committed Muslims makes the thugs untouchable to the police. Nowadays, at least in Eris's area, it is only the establishments who have not paid their protection money that incur the wrath of the 'zealots'.

But in fact, we have found that if you ask the correct question correctly and you get the right person to do the asking, you can get quite reliable information.

Again, there are manuals to guide anyone wishing to make a survey of sexual behaviour.[3] All dwell on the difficulty of asking 'sensitive questions' about sex and drug-taking. And indeed, if I were to ask you whether you prefer to be the insertive or receptive partner in anal sex, or whether you've had a foul-smelling vaginal discharge in the last year, you might think I was getting a little personal. But not everyone feels that way.

Don't forget that a person is being asked these questions because they sell sex, or come from a population likely to buy sex, or were recruited in a bar where men go to pick up other men. The research teams make no secret of that; once we've told the respondent we want to talk to them precisely because of their risky behaviour, there is no reason to be coy.

When I ask Nana how many guys she sold sex to last week, it's like asking my plumber how many leaks he has fixed – it is just getting people to talk about their work, and most people are happy to do that. When Desi was indulging her preference for heroin, she shot up around three times a day. She knows I know she's a junkie. Why would she tell me once a day, or six times? It's less effort to tell the truth.

When things go wobbly in the data, it is often because we have asked the question incorrectly. Even though she spent a lot of the time high, it is likely that Desi could remember how many times she injected the previous day. But if we ask about how many times she injected in the last week, or worse still in the last month, she's going to get pretty vague. And if you think she can remember how many times she used a new needle for an injection in that time . . .

People have a hard time remembering details about very

common events. Perhaps you, like me, regularly drink wine. Can you really remember exactly how many glasses of wine you drank in the last week, or the last month? And if you are asked how many of them were red wine and how many white, can you recall? Or do you shuffle through your mind and say, 'I guess I drink wine about three times a week, usually about two glasses. There was that bender last Saturday, but that wasn't typical, so I won't count it. Since it's winter I'm now drinking mostly red. So in the last month that would be four times three times two glasses of wine, twenty-four, and probably about eighteen of those were red wine.' You might even go on to think, 'Twenty-four glasses a month: does that make me an alcoholic? Since I'm only guesstimating anyway, I'll just say twenty and be done with it.'

Is that guesstimate 'accurate' enough? It depends to a certain extent how you are going to use the information. In routine surveillance of risk behaviour, we tend to care more about how things change over time than we do about absolute levels of risk. As long as people 'guesstimate' in the same way year after year, we'll still see changes over time. If a year from now you've cut back your drinking and estimate that you only drink on average twice a week, all of it red (because you've read that red wine reduces the likelihood of heart attacks), you'll calculate around sixteen glasses a month, all red. The information might suggest that the trends are going in the right direction (to someone who runs a campaign against alcohol abuse) or the wrong direction (to someone from the Red Wine Marketing Association). Overall, the guesstimate is good enough for our purposes, but we know it is just that, a guesstimate.

If I ask what you drank yesterday, you're almost sure to remember, and unlikely to lie. For some reason, people tend to be more truthful about a specific event (What did you do yesterday?) than about habits (What do you do most days?). Asking about what happened on a single day will not tell me much about your drinking habits as an individual – I might chance upon a day when you were hung over and drank nothing but coffee, but I can't conclude from that that you never drink alcohol. But in public health surveillance, we care less about individuals than we do about populations. And if we ask several thousand people what they drank yesterday, we'll get a pretty accurate 'snapshot' of overall levels of drinking in the population.

Chat with guys who like to buy sex, and you'll often find that they go to sex workers the way I go for shiatsu massage: once a week unless I'm travelling – though for guys buying sex it tends to be reversed: 'once a week unless I'm at home with the wife'. Being at home with the wife is the exception in the populations we're talking to – sailors and truck drivers and soldiers. But when you ask these clients how many times they have been to a prostitute in the last year, they'll tell you six or seven times on average. The numbers just don't add up; our snapshot of commercial sex is out of focus.

Out of focus is bad. To plan HIV prevention services for clients of prostitutes, we need an idea of how many men buy sex. We calculate this in steps. We start by estimating the number of prostitutes through a combination of mapping locations where sex is sold and a careful census of selected areas. Once we know how many women sell sex, we look at how many clients they have in a week, and how many weeks they work in a year. From that, we can estimate the number of transactions. Commercial sex is a zero sum game: every time someone sells, someone else buys. So the total number of clients in a year must be the number of sales, divided by the average number of times that each client buys in a year.[4]

If each client is responsible for a large number of the transactions, then the estimated number of clients will be lower. If each client buys only a few times a year, you will need a large number of men to account for all those sales.

In Indonesia, with men buying sex just six or seven times a year, this method gave us an estimate of around eight million men buying sex every year, one in seven of the adult population.[5] That is a high proportion by any standards; I was thrilled. The Health Ministry had started to take AIDS a bit more seriously – that was why they had decided to replace the back-of-the-envelope estimates made in 2001 with a more rigorous effort. But we were still having trouble convincing colleagues in other parts of government – places like the Finance Ministry, for example. I knew from Geneva how effective the 'innocent wives and babies' argument could be. And here we had eight million men buying sex. Their philandering could have serious knock-on effects, potentially exposing millions of wives and children to HIV, we argued. It was a total beat-up, of course. It doesn't matter how many millions of clients there are. They can only expose their wife to HIV if they have both a wife (which many did) and HIV (which most didn't). In fact, the estimates showed that for the time being only around 16,000 women were potentially at risk because of their husband's bad behaviour, but we didn't let the facts get in the way of a good story. We just *implied* millions of men buying sex equalled millions of wives at risk. The numbers were presented enthusiastically to the national cabinet, and much bandied about in the press.

The huge number of clients was actually implausible, because it came from an implausibly low number of visits to prostitutes each year. We realized that the numbers made no sense after we'd collected data in just three provinces. So we added a question to the survey in the remaining provinces: how many times have you bought sex in the last month? Sure enough, nearly everyone who had bought sex in the last year also said they had bought it in the last month. Twice in the last month, in fact. In other words, buying sex is a habit – like shiatsu, or going to a yoga class or reading fashion magazines – that people tend to do regularly or not at all. Very few people buy sex only once in a while. That seems to be true, at any rate, of the sailors, truck drivers, soldiers and taxi drivers who were in the surveys in Indonesia, and we see similar patterns in other countries. But if that is true, we'd expect men to report an average of at least twenty-two visits to sex workers a year – two a month for eleven months (with a month 'off' for the Ramadan fast). So why did men say they only bought sex six or seven times in the last year?

Actually, they didn't. More than a quarter of the 5,000 clients we spoke to said they couldn't remember how many times they had been to a prostitute in the last year. Not knowing what to do with these 'don't remembers', we just dropped them when calculating the average number of visits in a year.

But almost everybody could remember how many sex workers they went to in the last month, and perhaps predictably, men who couldn't remember their annual total reported far more visits in the last month than the rest. Again, it is that 'the more you do it, the harder it is to remember' thing. Ask me how often I went to a medical service last year, and I can tell you precisely: once. Ask my friend who has breast cancer, and I guarantee she won't remember. It is just too often.

In one study, if men answered 'don't remember' when we asked them how many partners they had in the last year, we asked them a follow-on question. Was it more than one? More than five? Ten or more? Almost all of the original 'don't remembers' were in the 'ten or more' category. If we just ignore the 'don't remembers' when we try to use the information to plan HIV prevention, we are going to be ignoring precisely the people with the riskiest behaviour, so we'll underestimate the danger of HIV spreading through commercial sex. But we'll also overestimate the total number of clients, so we'll overestimate the danger of HIV spreading to wives and infants. In short, we'll plan the wrong HIV prevention programmes.

So we have another two points for the honesty box. One: we don't always get reliable answers because we don't always ask sensible questions. We ask people to remember things they can't remember, or to calculate things they can't calculate. Sometimes, we ask the wrong questions simply because we are trying to collect information for some arbitrary 'global' indicator fixed upon by some committee in Geneva. They want us to report 'Per cent of injecting drug users never sharing equipment in the last month', so we ask injectors to remember something most can't possibly remember.* Two: we sometimes 'fix' the data before we use it, to try to make up for having asked the wrong question. After we added the question on visits to sex workers in the last month, we recalculated the number of clients, using that number times eleven to get an annual total. There were a few who didn't remember their monthly total, so we put them at the level of the top fifth of those who could remember. The estimated number of clients fell from eight million to 2.9 million.

* This indicator comes from 'National AIDS Programmes: A guide to Monitoring and Evaluation', a UNAIDS publication which I wrote in my Geneva days in conjunction with countless consultative committees, before I had ever spoken to a junkie. When I arrived in Indonesia, my colleague Wayne Wiebel, who had been working with drug injectors for decades, pointed out the futility of asking injectors to remember a month's worth of behaviour, so we started asking about the last day and the last week instead.

Dying to please

Asking the right question is key. But getting the right person to do the asking is pretty important too.

It used to be held as an article of faith that 'peers' and NGO workers are best placed to ask about 'sensitive' and illegal behaviours. This can work, but it can also backfire pretty badly. Imagine someone comes around asking questions about your salary. Would you give an honest answer to your 'peer' – the guy who sits across the corridor from you at work, a guy who usually does the same job as you but has for the moment been given a clipboard and the authority of being on a research team? Or would you be more likely to tell the truth to some accountancy student you've never seen before, who has no special interest in what your salary is beyond filling in a form?

An NGO worker in Burma told me he used to train sex workers as researchers, but it didn't go down well with the girls being interviewed. 'They'd say: she's a whore just like me. Why should I tell her anything?' he said. Now he puts all of his research team in white coats, with medical accessories. The stethoscopes do the trick. 'If they think you're a doctor, they'll pour their hearts out.'

Another difficulty of using NGO staff or other HIV prevention workers as interviewers is that they have a vested interest in the result of the study. Meet Al, who works for Pelangi (Rainbow), an NGO that is trying to keep the lid on HIV infection among gay men and rent boys in Jakarta. He hangs out in the same sorts of places as my love-interest Bhimanto, places which change as quickly as a gay man's wardrobe. A while back, he used to cruise Prego, the very bar where Bhim came out to me. But Al was not there for fun. Well, not only for fun. Al's job is to trawl the gay bars, nightclubs and massage parlours of

Jakarta, ladling out charm, advice on how to have disease-free fun, and safe sex packs. Safe sex packs contain condoms to keep body fluids with their owners and lubricant to reduce friction, keep condoms from bursting and keep body tissues intact. Plus information on HIV to encourage people to use the condoms and the lube. The packs come in three designs. One has a head-shot of a cute boy with come-to-bed eyes, another a torso with washboard stomach, a third a pair of low-slung jeans. Collect all three and you've got the Whole Boy.*

The job suits Al well; he would be in the clubs anyway, and this gets him in free. Often, though it is definitely not part of the job description, you'd find him on the dance floor, eyeing up some well-preserved Western guy who was in turn eyeing up some local talent even younger than Al. But quite a lot of the time, at least when I was around, Al would be doing what he was paid to do: plopping down on the brown suede sofa in the corner of Prego and talking to any guy who's there. Because he's so cute he was never turned down; people would listen to him talk about the importance of HIV testing just for the pleasure of listening to him talk. Bags of charm he was born with; bags of safe sex packs are provided by the Ministry of Health and FHI.

Al knew everyone and could get in everywhere, and that also made him a natural choice as part of a research team. But here's the rub: Indonesia was doing behavioural surveillance in large part to assess whether prevention programmes worked. So Al was assessing whether he and his colleagues have had any effect on the behaviour of gay men in Jakarta. If the results showed that lots of gay men had been reached with information and safe

* You can see the Whole Boy in the Gallery section of www.wisdomofwhores.com.

sex packs, and better yet if condom use and HIV testing had risen sharply, then Al's NGO would get more money, and he could keep on doing this job for another year.

Did that affect his work on the research team? It probably wouldn't make him tick the 'yes' box when the respondent in fact said, 'Condoms? Blerch! Never touch the things.' But it might well have affected who he approached with his clipboard. Why choose someone you've never seen before if you could choose someone you referred for HIV testing just last week?

Now think of it from the other side of the clipboard. Al comes up to you with his melting smile and asks if he can have a few minutes of your time. You melt. You've seen him around, in his Rainbow T-shirt, and you know he carries a bag from which condoms and lubricant magically appear. Like pretty much every guy in Prego, you know perfectly well that you should be using condoms when you have anal sex. And you do, most of the time. Well, sometimes. Anyway, you always mean to. And you know Al wants you to. So when he comes to asking you whether you used a condom the last time you had sex, you say yes. Because after all, if you had had one handy, you probably would have used it.

The tell-the-interviewer-what-you-think-they-want-to-hear lie is so common that it has a name in the epi-jargon: 'desirability bias'.

We began to wonder if it might be better to use outsiders to ask about sex and drugs. Preferably, outsiders who already knew how surveys work, outsiders who would fill in questionnaires correctly and hand them over diligently to supervisors who would in turn feed them into a system that would eventually spit out nice clean data sets. This was the thinking behind using Indonesia's National Bureau of Statistics to run behavioural surveys for HIV.

First reaction: you can't be serious. You can't ask boring middle class civil servants to go and do surveys in brothels. But boring middle class civil servants, especially ones who are often away from home doing different surveys in different cities, turn out to have the inside track on red light districts. Dudi, a colleague at the bureau of statistics who twinkles with enthusiasm about almost everything, laughingly calls himself the 'Mayor of Dolly'. Dolly is aptly named; near the port in Indonesia's second city of Surabaya, it is one of the largest red light districts in South-east Asia. In block after block, alleyways are crowded with bars and massage parlours, all pumping out their dangdut music. Dudi had been posted in Surabaya for years. Unlike interviewers from NGOs or branches of government that ran HIV prevention programmes, Dudi's job didn't depend on the outcome of the surveys, and he therefore had no incentive to influence the results, even subconsciously. To that extent, he was an outsider. But he certainly knew a lot of people in Dolly; in Surabaya, the mapping was done in record time.

Not all civil servants know the sex industry as well as Dudi. But with a bit of training, most seemed able to morph from wide-eyed innocent to seasoned observer in remarkably short order. I watched it happen in the most unlikely places, places like Merauke – a sad, dusty town stuffed into the bottom right-hand corner of the Indonesian province of Papua. Native Papuans are Melanesian; they differ from most other Indonesians in just about every respect: skin colour, facial features, diet, religion, language. But the majority of the population in Merauke is non-Papuan Indonesian, often from the Indonesian heartland of Java. They are referred to as 'newcomers', even when their families have been in Papua for generations.

Papua is at or close to the bottom of the Indonesian league table in almost all the major indicators of health and human development. It is also the wealthiest province, with vast resources of gold, copper, natural gas and tropical hardwood. For decades, the wealth was siphoned off by the central government in distant Jakarta, 2,300 miles away. But following a partial autonomy deal in 2002, Papua has been allowed to keep most of its revenue. Suddenly, the poorest part of Indonesia is awash with cash, and entertainment is king. Even the brand new Mitsubishi Pajeros that serve as taxis have built-in videos, so you can watch the graceless bump and grind of dangdut singers as you bump and grind over appalling roads – the wealth hasn't quite trickled down into the infrastructure yet.

When I was first there in 2003, the 'good' jobs hadn't trickled down, either. A lot of Papuan men were making very good money working in the mines and the other extractive industries; they were flooding the bars and brothels on their days off. But the white collar jobs were still dominated by 'newcomers'. These jobs, which actually pay far less than mining jobs, are coveted because of the respectability, pension and permanence they bring. The few Papuans who have managed to snag a civil service job are not in the bars; they tend to be deeply conservative, deeply Christian and permanently on best behaviour.

One of these young gentlemen saw his best behaviour dissolve when his ministry assigned him to work on the stats bureau's HIV survey team. He had been slow in the training: his nose twitched like a nervous rabbit's whenever he had to use the word 'sex'. In a twelve-page questionnaire about sexual behaviour, that's a lot of twitching. The minute we propelled him into a karaoke bar, he became a rabbit in the strobe lights. The bar was run by Andi, a well-built army corporal who is remarkably solicitous of the welfare of his staff. It looks much like any other karaoke bar in a third-rate town in a far-flung Asian province. The ambient lighting is an unflattering pink neon, supplemented over the dance floor by multi-coloured strobe lights, which bounce off a classic 1970s disco ball. The dance floor doubles as a stage. A huge flat-screen TV and a towering sound system take up a good chunk of the dancing real estate. These are the pulsing lungs of the karaoke club, the source of agonizing laments sung tunelessly in a scrambled language while generic blonde couples waft on video through flowered meadows. Occasionally, a beach does duty for the meadow, and a tearful Japanese teenager substitutes for the wafting blondes, but the agonizing laments are more or less standard.

I could understand the discomfort of my friend the rabbit. I myself have a morbid dread of karaoke bars, springing perhaps from the time that drinkers, fearing for their ears, actually unplugged the microphone into which I was 'singing'. In Andi's bar, though, no one was yet drunk enough to be singing. For now, couples and small groups were still dotted around the club's many dark recesses. Clients of every hue were installed on gold velveteen banquettes. Across them were draped the limbs of light-skinned young women who tugged absently at their pussy-pelmets while urging their clients to buy more overpriced drinks. These girls were imported from North Sulawesi, 1,200 miles to the west, an area known for the beauty of its (mostly Christian) women, and the exoticism of its cuisine – roast dog and bat stew are among the local specialities.

They will work for six months here in Merauke, the back end of beyond, then move on. With all the money now sloshing around in the local economy, prostitutes in Merauke earn among the highest fees in Indonesia – over half a million rupiah or fifty dollars a shot, two month's wages for an industrial worker in Sulawesi. The girls want to make the most of their time here by luring as many men as possible into the grotty clapboard rooms in the courtyard behind the bar, where the real business is done.

Young Rabbit Nose had never seen such a feast of wickedness. Under the watchful eye of the supervisor, a no-nonsense matron from a women's development NGO, he began his first interview. Still no singing but the music was loud, and he had to get very close to his target – a girl of about seventeen in a spray-on gold lamé top and a microscopic skirt. His knees were knocking against hers, and that sent his clipboard bouncing in all directions. He was trying hard to keep his eyes on the questionnaire but they kept straying to her generous cleavage. In his nervousness, he read out the instructions meant for interviewers as if they were part of the question: 'Circle 1 if answer is unprompted, 2 if prompted.' When he got to the questions about condoms, he broke into a lather of embarrassment, and the supervisor stepped in to take over.

Because this was a field test, the data wouldn't be used anyway. The whole point of doing field tests is to spot problems like this and sort them out. The statistics bureau manager arranged for extra training for this poor chap, and offered him a transfer to other duties. But he was hooked. By the end of the first week he was bantering back and forth with the girls, and I heard later that he had become a habitué of Merauke's bar scene. A veritable Rake's Progress, courtesy of the national public health surveillance unit. Upsetting, perhaps, for his nearest and dearest, but good for us. As the manuals stress, the key to good interview technique is to be comfortable with the respondents.

Another thing for the honesty box, then. People give different answers, depending on who is asking them the question. Neutral interviewers tend to be good for the data but bad for the budget. They need a bit of extra training before you let them loose in a world they have never dared to enter to ask about behaviours they had never imagined existed.

The Indonesian experience has shown that statistics bureaus can manage complex risk behaviour systems exceptionally well. But we also found that it works best of all when they co-opt 'peers', people like Lenny and Desi and Al, not to act as interviewers but to help with mapping, questionnaire design and field supervision. These people are good, too, at explaining to civil servants who does what to whom in anal sex, or what word to use to distinguish a regular client from a boyfriend. This model has made for some unlikely friendships. It has been fun to watch stuffy civil servants loosening up around blatantly gay co-workers, just as it has been fun seeing screamingly camp rent boys adopting a new 'grown-up' persona as they give presentations at meetings in government departments.

There were bits and pieces piling up in our honesty box, but we had been able to fix a lot of the problems as we went along. We got better at reaching the right people, and at training the right people to ask the right questions. As the data started to come in, there was encouraging evidence that people were telling the truth. Clients reported paying just about the same as prostitutes reported earning, with very similar geographical variations. Close to 7,000 sex workers were asked if they were carrying a condom, and two-thirds said they were. 'Show us,' demanded the interviewers. Polite to a fault, the prostitutes dug into their purses or pockets and brought out condoms. Fewer than one in five was lying.

Things were beginning to feel a bit more under control. But there were still many steps to go before we had all the information we needed to plan good HIV prevention and care programmes for Indonesia. And that was why we were going to all this trouble in the first place.

Code red

Once the information has been collected, we have to get it into computers, check it for errors, code it and analyse it.[6] Data management is mind-numbingly dull, but without it we would simply not be able to use any of the information that we have so painstakingly collected. So some poor bastard has to decipher the ticks, circles, numbers and words, often smudgily written by interviewers who are working at midnight in a place with crashingly loud music and romantically low light, and type them into a computer.

This job is performed by people who are thinking about their new girlfriend or their mother's birthday present or whether they can get twenty questionnaires entered and still get home to cook supper for the kids. Mistakes are common. So common, in fact, that most good research organizations get every data set entered by two different people, so that they can compare the two records and spot any inconsistencies that may be the result of a data entry error. This comes on top of the use of special software which flags potential inconsistencies and errors, so that you can't get away with entering a report of four husbands for a respondent who has never been married. Checking for these errors and fixing them is what we call 'cleaning' the data.

One reason errors are common is that statistical software does not deal well with words, so we tend to 'code' our data, giving each answer a number: 01 means primary education or less, 02 means middle school, 03 high school, etc. Getting the codes wrong leads to results that are always puzzling and occasionally catastrophic. Earlier on, when I worried about switching Pink T-shirt's study code with Nose Stud's blood samples, I was thinking about implausible study results. But what if you are giving people their HIV test results and you mess up the study numbers or miscode negatives as positives?

Sloppy coding can throw spanners into much larger works, as well. In 2001 a group of epidemiologists gathered in Melbourne to review the latest regional data ahead of the Asian AIDS conference. Fonny Silfanus, one of my many wonderful colleagues in the Indonesian Ministry of Health, reported new data from female sex workers in the massage parlours of Jakarta. HIV prevalence had suddenly shot up from virtually nothing to 18 per cent. That's one in six sex workers infected with a fatal, sexually transmitted infection in the capital of the world's fourth most populous nation, the city where I lived. When I first saw the data, you could have knocked me down with a Durex Fetherlite™.

That very week, WHO's Asia offices were releasing a report saying that HIV epidemics among drug injectors were unlikely to have any influence on a wider epidemic.[7] A number of us thought Asian politicians would interpret this as 'you can forget about the junkies, they'll kill one another but no one else', so we were looking hard for evidence of a 'cross-over' from injectors into other groups. The Jakarta data seemed to confirm that an epidemic previously confined largely to drug injectors was 'taking off' among heterosexual prostitutes and their clients. And after clients came their wives and children – the spectre of an epidemic in the 'general population'. Hurrah.

Dr Fonny called her colleagues in the national AIDS programme in Jakarta to double-check. Yes, 18 per cent, code 04 (which was massage-parlour-based sex workers), central Jakarta. Let's just call the lab and see if something could have gone wrong there? No mistake, they say: 251 tests, 44 HIV positive: 17.53 per cent. What was the expiry date on the test kits? A new batch, expiry early 2002. It all seemed to check out. By the time the Melbourne AIDS conference started, we'd written and released the 'Monitoring the AIDS Pandemic' report, containing a high-profile section entitled 'Indonesia – HIV takes off after years of silence'.[8] The Indonesian Minister of Health, who was at the conference, pledged action.

It wasn't until we got back to Jakarta and went down to the city health office to discuss the results that the truth emerged. We'd been tripped up by Indonesia's headlong dash for decentralization. Encouraged by the World Bank, Indonesia had recently smashed the centre's grip on the country, and had scattered responsibility for health, education, taxation, investment policy – more or less everything except defence and foreign policy – to the country's 300-plus districts to deal with.* The districts joyously started doing things their own way, with predictably chaotic results.[9] In this case, Jakarta city health workers had started using their own code book, rather than the national standard. Their code 04 was our code 05. The 18 per cent infection was not among sex workers, but among prisoners at Salemba prison, temporary home to hundreds of male criminals, drug dealers and injectors. The fact that one in six is infected with HIV was still hideous news – a year earlier not a single one of the 498 prisoners tested had been infected. But drug injection is an effective way of spreading HIV, whereas sex is not. The results remain shocking, but they are no longer surprising. They certainly didn't support the 'HIV has spread from drug injectors to prostitutes and your husband will be next' line that we were taking.

* Throughout the early 2000s, the number of districts changed almost weekly, growing from 325 in 2000 to 416 in 2003. By 2007, there were over 460 districts and municipalities, according to the national statistics bureau. However, districts that exist on paper often take some time to become functional in practice.

I'd like to put coding errors in the honesty box, but in fact they rarely get discovered. We found the prisoner–prostitute switch only because the results were so shocking. Even when coding mistakes are discovered, they can be hard to exorcise. In this case, the error was easily fixed in the national database, but it had already burrowed its way onto the internet and into the public record internationally. The ghosts of the 18 per cent of sex workers supposedly infected with HIV in Jakarta in 2001 have haunted the Indonesian AIDS programme ever since, raising the spectre of a heterosexual epidemic and channelling money away from drug injectors, where it is most needed.

Saints in white coats

You'll notice that in the heated phone calls between the conference in Melbourne and the lab in Jakarta no one even thought to question the coding. Our concern was about the HIV tests themselves – which test kits were used, the expiry date on them, that sort of thing. As a rule, however, very little attention is given to laboratory errors. Over and over again I get asked, 'How can we believe what the surveys say about sex and drugs? Surely everyone lies.' But no one ever asks, 'How can we believe the lab data?' People think chemical tests performed by nerds in white coats never lie.

In truth, the biological honesty box is at least as full as the behavioural honesty box. You have the same difficulty reaching the right people as you do in a behavioural survey. More difficulty in fact. People who don't mind answering questions about their sex lives often do mind you sticking a needle into them. And there's a slew of other things that can and do go wrong.

I wasn't totally blind to this when I arrived in Indonesia. When I was still commuting between Kenya and Geneva, the

Kenyan health ministry had asked for some help updating their national report on HIV. I went to see government epidemiologist Godfrey Balthazar, and he showed me the latest stats. In Nakuru, 55 per cent of pregnant women had tested positive for HIV. The suave Dr Balthazar mentioned the figure almost casually though it meant that more than half of the pregnant women in the nation's third largest city were infected with a disease that would kill them before their kids got out of primary school. As if the numbers were only that, numbers, not women with families, ambitions, secrets, futures. I tried to imagine a national health officer in an industrialized country just gliding over the fact that half the women in Chicago or Leeds or Osaka were infected with a preventable, fatal disease that would also be passed on to a third of their infants. I couldn't.

After some discussion Dr Balthazar agreed that the figure was astonishingly high, and that an error was likely. He phoned around to investigate, running through a checklist of things that might have gone wrong. Was it a very small sample size? (I've seen some health departments report 67 per cent HIV prevalence when they have only tested three people and two of them were positive.) But no, they had tested several hundred women. Were the chemicals used in testing out of date? No. Were the samples actually from sex workers rather than pregnant women? No.

Maybe it was true, then. The Kenyan government had borrowed US$40 million from the World Bank for HIV prevention just a couple of years earlier, but it could not find the money to pay for Dr Balthazar to drive the three hours from Nairobi to Nakuru to investigate this potential health fiasco. In the end, a British-funded HIV project financed his trip to the clinic where the samples had come from, and to the laboratory where they were tested.

The problem, it turned out, was that the laboratory ran out of sterile tips for pipettes, the little glass tubes used for dropping blood samples into the testing wells. So they used the same one several times. This meant that HIV from one sample could hang about in the pipette tip and get mixed up with the next sample. Bingo, HIV doubles. Again with money from the kindly British taxpayer, the surveillance round was repeated in Nakuru, now freshly furnished with a large supply of sterile pipette tips. People seemed relieved when HIV infection among pregnant women in Nakuru turned out to be 'only' 27 per cent.

Unlike behavioural surveys, which use little more than clipboards and pencils, biological testing requires things like test-tubes and pipettes, syringes and chemicals and laboratory equipment. That means a budget, and procurement procedures. Where there is procurement, there is the possibility of bribes and kick-backs. Money tends to distort science.

I came across an example soon after I arrived in Indonesia. The Indonesian Red Cross controls the nation's blood supply, and tests two million bags of blood a year to make sure they are free of HIV, hepatitis and other blood-borne pathogens. HIV surveillance activities add another few hundred thousand tests a year. All of the test kits are supplied by the Ministry of Health. They need to think carefully about which kits to buy, because different kits have different levels of accuracy.

With HIV testing, there are two things to worry about. If you are testing blood for transfusion, you want a test that is very sensitive. A very sensitive test will often give a positive result even if the blood does not have antibodies to HIV, but contains some other antibodies that confuse it. Blood serum that contains lots of antibodies to exotic pathogens is known in the trade as 'sticky serum', and blood from the outer reaches of the Indonesian archipelago is often sticky with yaws, enteric infections, rickettsial disease and other things you've never heard of. So you'll probably get a lot of false positive results with a sensitive test, but who cares? If you're screening blood before pumping it into a sick recipient, it is much better to be safe than sorry. You can just throw away anything that tests positive, and not worry about it.

If you're testing blood because you want to know how much HIV there is around – for surveillance, in other words – then you *do* care if there are lots of false positives. In this case, you want a test that will definitely give you a negative result if the blood is indeed uninfected – a quality known as specificity. You may fail to pick up one or two positive samples with a very specific test, but where HIV is comparatively rare, your results will be less distorted by false negatives than by false positives. And since you're going to throw the blood away as soon as you've tested it, it is no great drama if one or two positive samples escape notice.*

When I came to Jakarta, a debate was raging about which tests to use. The major consumer of test kits was the Indonesian Red Cross, which had been chaired since 1995 by President Suharto's eldest daughter Tutut. Health ministry officials told me that both they and the Red Cross had for years been under very heavy pressure to use a locally manufactured test kit. But now Indonesia was embarking on a new, democratic era, the Red Cross had a new chairman (the former reformist Finance Minister Marie Muhammad), and many different suppliers were wooing the government procurement officers, trying to sell their test kits.

* Here we are talking about screening anonymous blood samples. If an individual is being tested for diagnostic purposes, you would always use a minimum of two different tests, one sensitive and one specific, with a third as a tie breaker if you get different results.

I hadn't been in Indonesia long enough to understand who was selling which test kits to whom when I got a call from General Kiki Kilapong, a senior armed forces doctor. He wondered if we could help him with HIV surveillance in the Indonesian military. No problem, I said. I ran some numbers and estimated that we'd need a sample of around 4,000 to be able to track changes over time among the 57,000-strong uniformed forces. I wasn't sure he'd go for it – it seemed like a lot of soldiers having their blood taken, but when I went to see him with the spreadsheets he surprised me. 'Too many? No, I was thinking of 50,000.'

Fifty thousand HIV tests is not surveillance. It is case-finding – testing just about everyone in order to winkle out those who might be HIV-infected. I made grunting noises about the ethics of this. But Dr Kilapong insisted that this was not his intention. 'It's just that we've got all these test kits; they're expensive, and it would be a shame to waste them.'

The test kits were dropped on the army out of the blue, by a Ministry of Health that wanted them out of the warehouse, sharpish. I looked at the packaging: they were from some US company I had never heard of. As I walked back to my office, I ran into my boss Steve Wignall. No better person to ask: he had run the US Navy laboratory in Jakarta for years; a detailed knowledge of test kits is one of the many things he stores in the extraordinary hard drive that is his brain. Had he ever heard of Akers test kits?

'Oh God, not those snake oil salesmen from New Jersey?' A little shiver of distaste as he remembered seeing Akers executives make a sales pitch to Indonesian health ministry officials. Steve was heaven to work for. He was passionate about helping people on the margins of the mainstream, and had no time for people who leapt on to the HIV bandwagon just because they saw that the money was beginning to roll in. His greatest wrath was reserved for mealy-mouthed politicians who talked up their commitment to HIV prevention until the money came in, then lost their nerve at the first sign of voter distaste for effective programmes for injectors, gay guys and prostitutes. But he could be pretty unforgiving of the private sector too. 'Those tests kits are a piece of shit,' Steve said. His eyes narrowed with suspicion. 'You're not telling me they bought them?' I nodded, and Steve groaned, a deep-throated, why-would-I-have-expected-anything-else sort of groan.

Steve's opinion about the kits was derived in large part from an extensive comparative evaluation of test kits by Indonesia's national reference laboratory. The evaluation concluded that the Akers kits were neither sensitive nor specific. Oh, and hard to use. 'The Akers test sera have to be diluted up to 6 times to rid a pink discoloration of the background and to be able to read the results,' wrote lab guru Elizabeth Donegan, one of the authors of the evaluation, when I asked her about the kits. 'This can't be used in any field situation.' What's more, the kits have to be used on fresh serum, which means that testing has to be done where the blood is drawn – virtually impossible in a country the size of Indonesia where samples are usually frozen and transported to a provincial laboratory.[10] The lab specialist's dismal assessment was confirmed by colleagues in the health department in dusty Merauke. They told me that they sent their Akers positive blood samples to the regional reference lab for retesting using different kits. Eight out of eight 'positive' samples were actually HIV negative.

I tried to find out more from Akers about how the kits were tested, but got no response. All I could find on their website was a reference to a single study of fewer than 100 samples by a university in Nigeria. But as I trawled the web looking for information, I came across a six-page warning letter from the United States Food and Drug Administration addressed to Raymond F. Akers, Jr., PhD, President of Akers Laboratories. The letter (number 99-NWJ-03, dated 30 October 1998) is a matter of public record. It listed thirty-two separate ways in which the Akers test kit violates regulations for the manufacture, marketing and export of medical equipment, though it noted that the litany 'is not intended to be an all-inclusive list of deficiencies at your facility'.[11]

Why these kits were so attractive to the Indonesian officials charged with buying HIV tests is anyone's guess. But they were not attractive to the dedicated Red Cross staff charged with making sure that the country's blood supply was safe. They rebelled and refused to use them, according to a senior Red Cross official; that is why 50,000 more or less useless test kits got dumped on the military. I don't know if the military went ahead with the survey. If they did it was without our help – certainly the data never made it into the national surveillance system.

On the road

Of course laboratory equipment and the chemicals used in testing only become an issue if you have already got your specimens to the lab in good condition. That means keeping the specimens cool until they can be frozen, and then keeping them frozen until they can be tested. Some pointers for anyone planning research in tropical countries with erratic electricity supplies: don't forget to put diesel for the generator in the budget. And be nice to the people who live around the lab, so that they don't get annoyed with the generator noise and come and turn it off in the middle of the night. And make up a rota for supervisors to check that the generator is still powering the freezer rather than a local karaoke machine. All lessons learned at some cost to my equanimity.

My own worst moment was during a study of gay men in Jakarta. We were in Moonlight, one of Jakarta's grubbier gay discos. The place was packed, and the eternally cheerful club manager Eris had very kindly let us string my grandmother's best damask tablecloths across her office and the booze storeroom. This had created a warren of cubicles into which we inveigled survey respondents – bed-headed boys in sweat-soaked T-shirts who would rather be back on the dance floor.

The team squatted on upturned beer crates under a portrait of Megawati Sukarnoputri, the housewife-turned-president who would have found the very existence of Moonlight hard to imagine. From the wall she watched stoically as we took blood and asked questions about insertive and receptive anal sex. The team tried valiantly not to be distracted by a gaggle of waria who were strapping themselves into satin ballgowns for a floor show starring the irrepressible Bobby.

'Fancy me? I bet you do!' giggled Bobby. She grabbed at the crotch of Bambang, a nurse in the dull beige uniform of the city health department who was taking blood from a skinny bottle-blonde boy. 'But I only go out with ministers and celebrities, darling.' The long-suffering Bambang just grinned and got on with his work.

All in all, it was a long night. By about three in the morning we had forty-one blood samples on ice. The doof-doof music continued to throb around us, but most of the people still on the dance floor were by now too high to answer our questions. I decided to call it a night. The team packed up, and I strapped damask tablecloths, a cooler full of blood and a bag of used syringes to the back of my motorbike and set off for the laboratory.

Ahead loomed a police road block – a drug raid. Inconvenient. The cops want evidence of drug-related activity, and I am carrying forty-one used syringes. In a city where heroin is sometimes sold together with the means to take it, who but a drug dealer would have such a thing?

To make matters worse, these were not just any cops. The most reliable guide to a policeman's rank in Indonesia is not his epaulettes but his girth. The higher the rank, the better they are at extracting bribes. The better they are at extracting bribes, the larger the girth. Here in front of me I had Tweedledum and Tweedledee. They were so fixated on my cache of bloody syringes that they barely registered any surprise that a small white woman should be driving a motorbike through the entrails of Jakarta's 'entertainment district' in the small hours of the morning.

Like everything in Indonesia, this can be resolved with a dash of cash, which I keep clipped to the back of my driving licence for just such occasions. I reached for my wallet. But damnation, not content to deprive me of sleep, fill my lungs with smoke and ravage my ears with techno, Jakarta's gay scene harboured a thief who had swiped my wallet. Tweedledum and Tweedledee didn't mind: they'd impound my cooler as 'evidence', they said, and I could pay my 'fine' when I picked it up the next day.

By which time, of course, the blood would be ruined. That would be about an eighth of the entire study down the drain – and I didn't think the team could face another session at the Moonlight. I was limp with exhaustion and could feel my resolve sapping. Perhaps I should just give them the cooler and drive off. But the boys in Moonlight are special; they are cops and robbers, students and street vendors. They are not like the well-heeled guys who go to Prego and most of the other gay bars. If I ditched these samples, the study results could be badly skewed.

One last try. 'Of course, officer, I do understand. You're just doing your job.' I handed over the cooler, smiling sweetly. 'Oh, here, you'd better take some of these, too.' I fished out a handful of latex gloves. 'You just can't be too careful with HIV-infected blood.' The expression on those sleek-headed cops' faces imploded from self-satisfaction to horror. They thrust the poisonous blood back at me. Twenty minutes later the samples were at the lab, and I could go back to worrying about whether they would be tested using reliable kits and clean pipette tips.

Later, I checked the results of the Moonlight samples. To my surprise and relief, only one of the forty-one blood samples tested HIV positive.[12]

Is there still room in the honesty box? We need to squeeze in these issues: for biological results to be valid you have to handle your specimens properly. You have to use decent tests that have been stored properly. You have to follow the testing instructions correctly and use the right equipment. Slip up on any of these things, and your data slip, too.

Getting shirty

It was a pleasure working with the health ministry and the stats bureau; together with two first-class epidemiologists from Australia we made a happy band. The group worked hard at improving surveillance in Indonesia, reviewing the existing system, rewriting the national guidelines for HIV surveillance and developing new guidelines for behavioural surveillance. We drew new groups into the system – gay men, rent boys, waria – beefed up surveillance among drug injectors and collected a lot more information about how various groups interacted. As the system got stronger and the data improved, the health ministry pioneered ways to estimate the number of people who injected drugs or bought and sold sex as well as the number of men who cruised for sex with other men – work that was written up in the UNAIDS 'Best Practice' collection to provide a 'how-to' example for other countries.[13] Eventually, Indonesia developed one of the stronger HIV surveillance systems in Asia.

I hoped, even believed, that I had contributed to this. We were working our backsides off, we were wearing ourselves out, we were neglecting friends, partners and families. It paid off, because we were getting better and better data. But I was acutely aware that we still weren't translating all this information into services for the five million Indonesians at high risk of HIV infection.

The first time I ever went to Rawa Malang, Nana's swamp-side workplace, I was almost assaulted by the official boss of the red light district, Lukman. I was tagging along with a team from the statistics bureau. But in former colonies a white face is often wrongly deemed to be in charge, and soon I found myself hauled up before Lukman. 'T-shirts!' he bellowed. 'That's all we ever get from you. T-shirts! Out, the lot of you!' It transpired that Rawa Malang had been subjected to behavioural surveys three times already, under the earlier system run by the University of Indonesia. The researchers had said the surveys were necessary to plan HIV prevention programmes and they had left T-shirts emblazoned with anti-AIDS slogans as a thank you to survey participants. But Lukman didn't want T-shirts for his girls, he wanted condoms and a health centre – regular screening and treatment for sexually transmitted infections at the very least.

We had a team of twenty interviewers ready to swarm across the complex to ask girls about their clients and their condom use, but Lukman was having none of it. We were just about to leave with our tails between our legs when one of Lukman's lesser goons came in to announce the death of Lukman's counterpart in a rival entertainment complex near by. There was a bit of confusion; then the Poobahs of prostitution dashed for their cars and zoomed off to the same-day burial required by Islam. We were free to get on with our work.

I have to say, though, that I felt slightly guilty. I asked my colleague Irene Sirait why we weren't supporting any prevention programmes in Rawa Malang. Irene was a fierce North Sumatran who worked on our commercial sex team, and my question provoked her to great indignation. She said that we had at one stage been working with a local NGO to provide health services for prostitutes in Rawa Malang but the group pulled out, complaining of harassment from another NGO, whose programmes were backed by Church World Services. Irene herself had been delicately threatened by the rival funder. 'CWS called me and asked me not to support programmes in Rawa Malang, Rawa Bebek and Kalijodo [three of the largest red light districts in North Jakarta], because the area "belonged" to them,' Irene told me in an e-mail. The Christian organization was teaching sex workers to bake cakes; apparently their expatriate boss objected to having programmes that provided health services for prostitutes in the same area.

Lukman, for his part, objected to us demanding that people give up their time, their secrets, their body fluids, in exchange for nothing more than T-shirts and baking lessons. I agreed with him more with every passing day. When I had taken on this job I had wanted to help map the landscapes of risk, measure how much risk there was and then work to reduce it. We were doing pretty well at the first two, but we were still failing miserably at the third.

We could blame some of this on the chaotic government decentralization process which had been wreaking havoc in Indonesia since 2001. Surveillance was going ahead, but we didn't always get the results. Districts forgot to send data to their provincial offices, or provinces asserted their independence from central control by refusing to pass information on to Jakarta. Records were kept on scraps of paper that got stuffed in a filing cabinet; when districts split amoeba-like, no one could decide which filing cabinet belonged to whom. Each new district needed new staff; the smartest and best-trained got promoted out of the lowly surveillance unit taking their institutional memory with them. Provincial staff were supposed to train new recruits for the districts, but there was little expertise at that level, and no training budget. Many districts were staffed by people who didn't know the difference between surveillance (the anonymous testing of several hundred people in order to track trends in HIV over time) and diagnosis (a single test ordered by a doctor who suspected a patient might be infected with the virus). They reported 'surveillance' in a sample of two pregnant women, and invented the infamous 'etc.' category. At the central level in Jakarta, competent staff tried to hold the scraps together. They diligently entered what data they had into Excel spreadsheets, occasionally spotting rises in infection that had gone completely unnoticed in the districts because records had not been kept from one year to the next. But these 'a-ha' moments were rare. No one was given the task of analysing the data, and most of the surveillance results were not examined in any detail. It was as if surveillance was an activity in its own right; a slightly inconvenient annual procedure that just had to be ticked off, like signing up for the electoral register or paying your road tax.

Over the years I was in Indonesia, my colleagues in the ministry worked really hard to change this. They had provided a lot of training for the provincial staff, in the hope that some of it would trickle down to the district level where the work is done (under decentralization rules the centre can't deal directly with the districts). They had developed new national guidelines, and my Bali-based colleague Brad Otto had backed the guidelines up with a wonderful software program that threw up warning signs if anything was amiss in the data. The system wouldn't let you enter a sample size of two, or 'etc.' as a surveillance category. It automatically generated graphs and drew attention to any rapid increases in infection rates with messages in unmissable red type: 'This is an unusually rapid rise!' But even this did not seem to be enough to cajole health staff into going to the politicians and demanding action. Brad is an Australian epidemiologist who sings regularly in a Balinese drag bar and takes almost anything in his stride. But even he was frustrated when provincial health officers entered their data into the wonderful software, saw the infection curves headed skywards, and did nothing to improve prevention programmes. 'Are you aware of a computer peripheral shaped like a large 2×4 that can whack people over their heads?' he asked me in a plaintive e-mail.

HIV infection had hit 18 per cent among Jakarta's prisoners in 2000, but by 2004 there were still no HIV prevention programmes in the city's prisons, let alone any care for infected prisoners. We'd tracked pathetically low levels of condom use in an easily accessible red light district since at least 1998, but by 2003 we'd provided nothing but T-shirts for the girls in the area. We realized in 2004 that the antibiotics prescribed for STIs in the national treatment guidelines were completely ineffective, but it took three years to change the guidelines and by mid-2007 there was still no way to buy affordable versions of the effective drugs in Indonesia. In short, we knew what needed to be done, but a lot of the time we were still failing to do it. How was it possible?

There are many reasons for the failure to turn good information into good HIV prevention programmes. The simplest is that the people who collect public health data are not the people who can act on it. The gulf between those who have the information and those who make the decisions gapes wide. In a country like Indonesia HIV surveillance staff are nobodies; they work on a stigmatized disease in an unimportant ministry – the smart people go into finance, public works or energy and mineral resources, the 'wet' ministries that control budgets in the mega-billions. Surveillance staff don't have access to policy-makers, and even if they did they wouldn't have the skills to convince them to take on difficult challenges like doing nice things for junkies. What's more, they don't have the incentive. Being a successful civil servant is all about not rocking the boat. Storming around asking your superiors to give out clean needles in prisons is definitely rocking the boat.

There are other, more complicated reasons why we fail to turn more information into less HIV. They involve ideology, politics, money and history. I'm going to begin my examination of these issues by looking for a moment at the world's greatest and most shameful monument to failed HIV prevention: the AIDS epidemic in Africa. In Africa we've made every mistake in the book. Now we're exporting those mistakes, threatening to trip up HIV prevention in the rest of the world, too.

4

The Naked Truth

If you talk about AIDS these days, most people think of Africa. Two-thirds of the people infected with HIV in the world live in the countries south of the Sahara. They are clustered especially in the south of the continent, and infection rates are high in East Africa too.* Everyone knows it, and lots of people want to do something about it: Irish pop star Bono and serial do-gooder Sir Bob Geldof are on the bandwagon, and they've roped in English footballer David Beckham, who flashes his AIDS credentials by promoting 'red' i-Pods; a fraction of the sale price goes to AIDS in Africa.

Africa is a giant, in-your-face failure for the HIV prevention industry. It might be even worse without our prevention efforts, though it is hard to see how it could be. On the prevention balance sheet, we're in the red to the tune of 45 million infections. In some countries, over 80 per cent of all adults will die of HIV.[1] A schoolgirl in South Africa is thirteen times more likely to be

* The countries of East and Southern Africa are home to 3 per cent of the world's population, and 40 per cent of the people living with HIV.

infected with HIV than a woman who sells sex for a living in China. A civil servant in Swaziland is forty times more likely to have HIV than a junkie in Australia.[2]

How did we get into this appalling situation?[3] In large part because most African politicians found it easier to watch hundreds of thousands of young adults die than to say what everyone was secretly thinking: HIV is spread by sex. Most HIV is in Africa. Ergo, Africans must have a lot of sex. There, I've said it.

Far from taking African leaders to task for the wilful neglect of a disease that was ravaging their people, the I'd-rather-die-than-be-called-racist international development types let them off the hook. They found other reasons to explain why AIDS is being spread by unpaid sex between men and women in Africa but not on any other continent. What does Africa have more of than other continents? Poverty and underdevelopment.

Smoke gets in your eyes

A headline in 'Confronting AIDS', a landmark World Bank publication on HIV which appeared in 1997, sums up the smokescreen we managed to create around sex. 'Poverty and Gender Inequality Spread AIDS', it declared.[4]

The maths behind this statement was explained by one of the report's authors, Mead Over, during a meeting of the nerdy-sounding 'UNAIDS Reference Group on HIV Estimates, Modelling and Projections'. These nerd meetings are one of the things I like best about my work. You get to hang out with other scientists, people who have jumped through hoops that make police road blocks in Jakarta look trivial. They've snatched thousands of defrosting blood samples from the jaws of EU customs officials who wanted to keep them until some trade dispute was resolved. They have been attacked by Zimbabwean villagers who believed the devil would arrive in a white Land Rover to take their blood for satanic rituals. They've overcome all manner of trials and tribulations to help increase our understanding of how people get HIV and what happens to their bodies, incomes and families once they are infected.

Every now and then, we get together in nice places to swap notes. We're free of the self-censorship that goes with politically correct scientific conferences; we can use rude language and ask rude questions. And we're usually keen to get beyond the PowerPoints and out into some nearby piazza, to carry on our discussions over a decent bottle of wine. It beats trying to keep the rats out of rural hospitals or taking anal swabs in the slums, or whatever it is that we're doing most of our working life.

The sun was shining on a Roman piazza outside the room where Mead was explaining his maths in 2000, and the crowd was getting restless. I looked at Mead's graphs.[5] They showed that the poorer a country is, the more HIV there is. It was especially true where there was a wide gap between the rich minority and the poor majority. If men are more literate than women, AIDS spreads. If there are more men in the cities than women, AIDS spreads. These and four other factors, including the proportion of the population that is Muslim, were squashed into a complicated formula which apparently explained HIV. None of the factors related directly to sex or drugs. Surprising, since in his earlier work he had urged governments to concentrate on 'core groups', the people most likely to have a lot of partners at once. Sex workers and their clients, in particular.

Now sex seemed to be out, and poverty and gender were in. All the epi-nerds in the room had one question. I stuck up my hand and asked it. 'Mead, where's the hard-on in your formula?'

The World Bank believes poverty and gender inequality

spread AIDS. I believe sex and drug injection spread AIDS. The very best rational, utility-maximizing calculations tend to get displaced by erections and addiction.

To be fair, Mead and others see poverty, inequality and education as things which determine what people do (or allow others to do) with their erections. People need to sell sex because they are poor. They need to buy it because they are migrant labourers a long way from their wives. Women can't use condoms because they don't have anywhere to put a condom, and don't dare ask their lover, husband or client to wear one. Men can't use condoms because, well, because.[6]

Yes, our social and economic circumstances shape decisions we make about all sorts of things in life, including sex. Sometimes they rob us of the power to make any decisions at all. But of all human activity, sex is among the least likely to fit neatly into the blueprint of rational decision-making favoured by economists. To quote my friend Claire in Istanbul, sex is about 'conquest, fantasy, projection, infatuation, mood, anger, vanity, love, pissing off your parents, the risk of getting caught, the pleasure of cuddling afterwards, the thrill of having a secret, feeling desirable, feeling like a man, feeling like a woman, bragging to your mates the next day, getting to see what someone looks like naked and a million-and-one other things'. When sex isn't fun, it is often lucrative, or part of a bargain which gives you access to something you want or need.

If HIV is spread by 'poverty and gender inequality', how come countries that have plenty of both, such as Bangladesh, have virtually no HIV? How come South Africa and Botswana, which have the highest female literacy and per capita incomes in Africa, are awash in HIV, while countries that score low on both – such as Guinea, Somalia, Mali and Sierra Leone – have epidemics that are negligible by comparison? How come in country after country across Africa itself, from Cameroon to Uganda to Zimbabwe and in a dozen other countries as well, HIV is lowest in the poorest households, and highest in the richest households? And how is it that in many countries, more educated women are more likely to be infected with HIV than women with no schooling?[7]

For all its cultural and political overtones, HIV is an infectious disease. Forgive me for thinking like an epidemiologist, but it seems to me that if we want to explain why there is more of it in one place than another, we should go back and take a look at the way it is spread.

Back to basics

The 'AIDS is spread by poverty' myth is just the latest in a long series. The first probably dates from the hysteria of the 'gay plague' years, when AIDS was said to be spread by party drugs. Mythmaking is so common in the AIDS industry that almost everyone has questions about the facts, questions that surface late at night, in bars, after a couple of drinks. 'OK, you're an epidemiologist. Is it true that I can't get HIV if I'm circumcised . . .?' So let's just run quickly through what we know about HIV.

We know that HIV hangs out in blood and genital fluids but not saliva. Saliva confuses people because you can do an HIV test on saliva. That's because most commercial tests look for antibodies to HIV, which survive in saliva, rather than the virus itself, which doesn't.

We also know that HIV is not actually all that infectious. Unlike other viruses such as hepatitis, HIV can't survive for long outside the body, and can't survive at all outside of body fluids. Of course if you take infected blood and pump it straight into someone else's body, you'll be almost guaranteed to generate a new infection. That's why it is so important to screen blood for HIV before a transfusion. Injections, too, can put the virus right where it needs to be. If someone comes at you with a syringe in a clinic or hospital where HIV prevalence is high, make sure the equipment is new or has been properly sterilized. If you're injecting heroin, steroids or even insulin, use your own needle. If you don't, you risk giving HIV an express ticket into your bloodstream.

Sexually, though, the virus has a much harder time. It can't pass through the body's own lines of defence – the mucus-covered surfaces that line most of the entry points to our bodies. To take up residence in someone else's body, HIV more or less needs to be invited in through an open door. And injections aside, there are not all that many things that can open a door. The main ones are small tears and lesions around our genitals (or in our anus, in the case of anal sex), and the presence of cells which are especially welcoming to HIV.

Forgive me for getting graphic, but as you probably know, sex can be a sticky business. The stickier the better. The more mucus there is, the less danger there is that the friction of sex will tear your inner fabric. A wet vagina is usually a pretty safe environment. This is one reason why foreplay is a good thing – stimulation and anticipation make a woman wet, and therefore safe. It is also a reason why HIV is so easily transmitted in forced or unwanted sex. Some women believe that men prefer sex 'hot and dry', so they soak up their juices using anything from toothpaste to fertilizer.* This may or may not increase

* This practice has been recorded most frequently in southern Africa, notably Zimbabwe and Zambia. See for example Civic and Wilson, 1996; Sandala et al., 1995. However it exists elsewhere, too: Halperin, 1999. In Indonesia, some women use a drying potion that originates from the East Javan island of Madura. Madurese women are famed for their sexual prowess.

pleasure, but it certainly increases the possibility of vaginal trauma, and that opens the door for HIV.

The anal passage does not get wet in anticipation of sex. On top of that, its lining is not very thick. It's not a great idea to have anal sex without using an artificial lubricant. For one thing, it hurts. But more importantly (from HIV's point of view), it is quite likely that you'll get scratched up inside, so if one of you is infected, you can pass HIV on quite easily to the other.[8] Don't assume this is only relevant to gay men. A landmark study of heterosexual couples by Isabelle de Vincenzi (the clafoutis chef from Geneva) found that women who reported anal sex were over five times more likely to pick up HIV from their infected partner than women who only had more conventional vaginal sex.[9]

The other things that poke holes in the body's natural barriers are other sexually transmitted infections, especially the ones such as syphilis and herpes which cause ulcers. These wounds, external in men, internal in women, provide safe passage for the virus, out of one body and/or into another. They have another drawback, too. The particular genius of HIV is that it attacks the very cells that are supposed to fight off disease. If you have another sexual infection the body's defence forces (white blood cells known as T cells) come in to attack it. Now your genitals are swarming with T cells, and T cells are especially susceptible to HIV infection. Another express ticket for HIV.[10]

Another warm welcome is provided by the Langerhans' cells which sit on the inside of the foreskin that covers the uncircumcised penis. Because circumcision lops off these cells, it closes one of the doors through which HIV might enter a man's body. It also toughens up the tip of the penis, so that lesions are less common. You can still get infected if you're circumcised, especially if you're having anal sex or other forms of sex that might be particularly abrasive. But HIV spreads much more slowly in countries or cultures where most men are circumcised.[11]

Another thing I get asked about a lot after a glass or two of wine is oral sex. Is it a risk for HIV or not? Because there's no HIV in saliva, it's pretty hard to pass HIV on to someone else by going down on them unless you've got bleeding gums. And it's hard to contract it that way, either, unless there's a way for infected ejaculate to get into your bloodstream. Again, mouth wounds or bleeding gums.[12] HIV prevention programmes don't like to tell you this, but really, if you go to the dentist regularly the likelihood of infection in oral sex is vanishingly small.

Without special treatment, about one in three of the women who have HIV while they are pregnant will pass HIV on to her infant either in the womb, when the child is born, or during breastfeeding. One of the things which determines whether the child will be infected or not is how much virus there is in the mother's blood. The amount of virus in your body fluids (what the doctors call the 'viral load'), is critical in sexual transmission, too. Viral load varies depending on how long you've been infected with HIV. When the virus first walks through an open door into a new body, it steals a march on the body's defences and multiplies like mad. Then the immune system swings into action, creating antibodies that attack the virus. Although the immune system is never able to kick HIV out entirely, because the virus sneaks off to hide inside healthy cells, it does bring it under some kind of control. The viral load comes way down, and usually chugs along at very low levels for several years, although other infections can cause it to spike. Sexually transmitted infections are especially good at pricking up the levels of HIV in semen and vaginal fluids, which makes onward transmission much more likely. This is another reason why it is a good idea to treat curable STIs.[13]

All the time you're infected, HIV is chipping away at the body's immune system. Eventually, it gets the upper hand. Armies of the virus come out of their hiding place in healthy cells, and, with no serious opposition, are able to reproduce at a great rate. The immune system gets crushed even further, and the amount of virus in the blood soars.

The timetable for all of this activity varies. In general, however, antibodies start being produced in earnest after two or three weeks. Since most commercial HIV tests are for antibodies rather than the virus itself, it is at this point that a person begins to test HIV positive. The early spike in viral load is brought under control by a body's natural defences in something between six weeks and six months. If you're broadly healthy, nothing much happens for eight or nine years, until the immune system starts to collapse. The amount of virus in the blood soars again. Now you're looking at AIDS; your body can't fight off other diseases and you start coming down with tuberculosis, pneumonia, toxoplasmosis or any number of other nasty diseases that feed on a weakened immune system. Unless you start taking a combination of antiretroviral medicines which slow the rate at which HIV stamps out copies of itself in your body, one of these diseases will kill you within a year or two.

So: HIV is transmitted when the viral load is high. Your viral load is high immediately after you get infected – a time when you may well be doing risky things. After all, you got infected for a reason. It is high again when you are sick and close to death, when you're probably not going to have much luck picking up new partners, even if you had the energy to try.

That means most people who pass HIV on sexually do so when they're newly infected – often within the first six months. In fact, a lot of HIV is probably transmitted before a person could even test positive for the virus.* Obviously, that's nearly impossible to measure. But one case was brought to light courtesy of the health and safety regulations in the US porn industry. Porn stars are tested for HIV regularly. One stud who tested negative on 17 March 2004 had unprotected on-camera sex with thirteen women in less than a month. Then he tested positive, and the health authorities went back and tested all his partners. We usually bandy about rather low transmission probabilities in heterosexual contact – as low as 1 in 300 for male-to-female sex. But three of the women became infected by this one man – an attack rate closer to 1 in 4. They had some fairly unusual sex – I'll leave you to guess what 'double anal' sex is – but it was still a far higher rate of transmission than one would normally expect. And it was almost certainly because our screen stud had a brand new infection, and his semen was seething with HIV.[14]

The virus will spread most quickly in situations where people have several sex partners on the go at any one time, because they will be able to pass the virus on to all of them in their highly infectious first few months. If on top of that you have lots of untreated STIs, which spike the viral load and open doors for genital fluids, you'll accelerate the pace. Genital herpes is particularly good at fanning the flames of HIV: over two-thirds of adults have herpes infection in many badly affected areas.[15] Transmission will go faster still if a lot of men are not circumcised.

* Injecting is a rather different story. Because it's such an efficient way of transmitting HIV, it probably works even when there's not all that much virus in the blood. The risk has not been quantified, but it is likely that you can get infected by sharing a needle with a junkie whose HIV levels are 'under control' relatively easily, compared with the risk of sexual transmission from the same person.

Spreading the net

All of Africa is poor. But not all of Africa has very high rates of HIV. HIV reaches very high levels only in areas where there are lots of simultaneous sexual partnerships, lots of untreated STIs and lots of uncircumcised men.

The circumcision and untreated STIs are easy to understand, and they are relatively easy to measure. Sexual partnerships are more difficult on both counts.[16] A lot of studies compare averages – average number of sex partners, average age at first sex – but averages can mask all sorts of different things. Sexually transmitted infections tend to build up a critical mass thanks to people whose sexual activity is way above average.[17] In any case it turns out that HIV cares less about numbers than about patterns – the virus spreads better through webs of partners than through chains.

Compare Simon and Caspar.

Simon has a wife at home in the city, and also has sex with his accountant, a woman whose own husband is a bit of a rake and is often away. But Simon's business takes him away quite a bit, too – he goes to visit his major suppliers once a month, and he also makes regular visits to the port to oversee shipping operations. Naturally, since he is a frequent visitor, he has girlfriends in each of those places. Oh, and there is the childhood sweetheart in the village. He sometimes has sex with her for old times' sake when he goes to visit his parents. Simon is a solid type, and has been pretty faithful to these five women for a decade or more.

Caspar, on the other hand, is an incurable romantic. He falls head over heels in love, and promises to be faithful for eternity. But he has a romantic's limited attention span. To Caspar, a year is an eternity, and he keeps his promise for that time. Then he drifts off in search of a new, more perfect woman. He soon finds her, and falls in love again. He stays with her for an eternity, and then drifts off in search of . . . you get the picture. Caspar is monogamous; he's always faithful to one woman. But over a decade he keeps the faith with ten women – twice the number that Simon clocks up.

Though romantic Caspar has twice as many partners, solid Simon will do far more to spread HIV if he becomes infected. This is because he sees all of his partners regularly, and so is likely to have sex with all of them soon after he contracts HIV, just when he is most infectious. If Caspar gets infected, chances are he'll stay with the woman who infected him for a while. By the time he moves on his viral load will have come down and he'll be much less likely to pass the virus on.

Of course neither man can pass on HIV unless they get infected first by one of their women. And that depends on who their women are having sex with. Romantic Caspar is giving his girlfriend all of his attention. For as long as that lasts, she's less likely to have the need, the desire or the opportunity to have sex with other men. Solid Simon's partners, on the other hand, get only occasional attention. They've got big gaps in their calendar when he's off travelling, or with one of his other women. They may also have gaps in their purses (if he's a source of cash), and in their emotional and sexual lives. If they fill these gaps with other lovers, including their own husbands, then they are more likely to get infected than Caspar's girlfriends. And so Simon is more likely to get infected, and more likely to pass HIV through his web of partners.

The data show that there are more Simons in Africa, more Caspars in the West.[18] On top of that, people in many parts of Africa start having sex younger than people in other developing countries and have more sex before marriage. There's also more sex between older men and younger women.[19]

As Noerine Kaleeba sees it, sexual networks (at least in her native Uganda) are a hangover from the days when polygamy was the norm. Noerine is a force of nature and an institution in the HIV *demi-monde* as well as in her home country. She is full of good-humoured frustration at the vagaries of her many jobs – HIV counsellor, international bureaucrat, mother, friend and carer for the thirty-eight remaining members of an AIDS-ravaged family. She knows a bit about polygamy because her own father had four wives and twenty-eight children. She knows a bit about extramarital sex, too. When Noerine was a teenager in Uganda she went out with a classic 'sugar daddy', a good-looking, married forty-year-old in a Mercedes Benz. HIV approves of such age differences because it gives the virus a way of infiltrating younger generations. If people only had sex with others the same age as themselves, then the HIV epidemic would die out, because eventually everyone who was already infected would be dead. Uninfected people would be having sex with other uninfected people. Noerine is well aware of this. But she also knows the 'do as I say, not as I do' message is a tough sell even with her own teenagers.

When I lived in Kenya, teenagers referred to sugar daddies as 'Three Cs boyfriends' – Cash, Car and Cell phone. A lot of them were looking for relationships with these sophisticated older men, men who would give them things that spotty, fumbling adolescents never could. Men who would make them feel like a woman and teach them about sex. Men who would buy them dinner or give them perfume. Men who may well give them HIV.

There's a near universal assumption that most of the women infected with HIV in Africa are 'innocent wives' infected by 'promiscuous husbands'. That was probably true early in the epidemic. HIV will always rise first in the people with the highest number of unprotected contacts with people who may be infected. Heterosexually that prize goes to sex workers in almost every country. The nations of East and Southern Africa were not exceptions to this rule. Prostitutes get HIV from clients, and pass it on to other clients, unless condoms bar the virus's way. Clients take the virus on to their other partners – they'll probably pass it on most quickly to the one they have sex with most often, and that's often their wife. True to stereotype so far. But if the innocent wife isn't having sex with anyone else, the virus has nowhere else to go.

This is demonstrably not the case in the countries of East or Southern Africa, nor in a number of countries in West Africa. Plenty of wives are having sex with people other than their husbands. And even higher proportions are having sex before they ever have a husband.[20] A lot of girls are infected by a Three Cs boyfriend well before marriage. In East and Southern Africa, we consistently see far higher rates of infection in young women than in young men. The data that really made my jaw drop came from a study in the general population in Carletonville, in South Africa. More than half of the women in their early twenties were infected with HIV. Among men the same age, infection rates were a third as high. In some other settings, teenage girls are seven times more likely to be infected than teenage boys.[21] Many of these young women will marry men a lot closer to their own age than their sugar daddy's. Although they've usually slept with the younger man before marriage, as a sort of audition, sex is often sporadic (she'd rather spend her time with the guy who's buying her perfume).[22] So the boy makes it to the marital bed uninfected. In nearly every African country where we have data, unmarried girls are more likely to be infected with HIV than unmarried men.[23] Once a man marries, though, he'll have sex with his wife a lot more frequently, and condoms, if they were ever part of the picture, will evaporate. If she's carrying Sugar Daddy's virus, she'll probably pass it on to her husband sooner or later.

Now that so many young women are infected with HIV before their wedding night in Southern Africa, getting married is one of the riskiest things an uninfected young man can do.

The 'innocent wives' storyline is also dented by new information about couples where one partner is infected and the other isn't ('discordant' couples, in epidemiological jargon). Wives are the infected partner in 62 per cent of discordant couples in Kenya and Ivory Coast, and in ten other African nations the proportion never sinks below a third.[24] One detailed study in Uganda showed that while women were the HIV-positive partner in half of HIV-discordant marriages, men were twice as likely to bring HIV into the marriage as women were. That's partly because men transmit HIV to women faster than women transmit it to men, so men who have brought HIV into the marriage are less likely to have a 'discordant' wife by the time a researcher comes along.[25] But still, it is clear that millions of African women have been infected with HIV while having sex with men who are not their husbands.

Women are more likely to report more than one regular sex partner on the go at any given time in the African countries where HIV is high than anywhere else in the world.[26] That's key if HIV is to spread through the population. In our Geneva beat-ups, we liked to imply that once 'innocent wives' were infected by their philandering or injecting husbands, HIV would storm through the bustling populations of India, China, Indonesia – anywhere we wanted to turn scarlet on the map. But it wasn't true. Asia and some other parts of the world have been dominated neither by the Caspar nor the Simon model. They have a third model of heterosexual behaviour, Yuni's model. Yuni is married, but he likes a bit of variety. He certainly can't sleep with his accountant, as Simon does. She's absolutely devoted to her husband. Yuni does have a 'girlfriend' in the capital – she works in a bar and gives him special attention (for a discount) whenever he visits. And he occasionally goes out carousing with his mates when he's travelling, and they end up at a brothel.

If Yuni is not using condoms, he could pick up HIV infection from his girlfriend or one of her colleagues. If he does get infected, he'll take the virus home, and there's a good chance that he'll infect his wife over the next couple of months. But she's not having sex with anyone else, so HIV hits a dead end.

So what is the chance of Yuni getting infected? It depends, of course, on whether the girls he buys sex from are infected, and when they got infected. And that depends on how long they've been on the game, how many clients they have, and what they do about sexually transmitted infections.

Some brothel-based prostitutes turn over as many as five or six clients a night, but that's exceptional. The hardest working girls on record in national surveillance data are in Thailand, Cambodia and some Indian states – they used to average around twenty clients a week before HIV prevention programmes shooed a lot of clients away from brothels. Contrast this with countries where there is very little HIV: in East Timor, just three clients a week. In the Philippines, only two. China and Indonesia, around five or six in a good week. With such low client numbers, a prostitute is less likely to encounter a client who is infected, and less likely to be sore and susceptible to infection when she does make a sale. Most of all, fewer clients means fewer people at high risk of becoming infected during the all-important early weeks of infection. Put the window for high infectiousness at a conservative three months, and do the sums. If her clients don't use condoms, a newly infected prostitute in India's commercial centre, Mumbai, will expose 240 guys to a high risk of contracting HIV. A girl infected at the same time in the Philippines will expose twenty-four clients to her high viral load.

In Asian countries where client numbers are low *and* most men are circumcised – Indonesia, the Philippines and Pakistan, for example – HIV doesn't stand much of a chance heterosexually, even if condoms are rare. Fewer than one in twenty prostitutes is infected in these countries. Where client turnover is high and foreskins are still in place, HIV does much better. The virus stormed through the commercial sex industries in Thailand and Cambodia, infecting over a third of prostitutes and hundreds of thousands of clients in the early and mid-1990s. Then condom use rose, and visits to the brothel fell. So did HIV.[27]

Think again about what it takes to transmit HIV. One: a person with lots of HIV in their blood or genital fluids must have sex or inject drugs with an uninfected person. Two: body fluids must pass from the infected person to the uninfected person. Three: the virus must find an 'open door' in the uninfected person. If you're HIV-free and want to stay that way, you have to make sure at least one of those things doesn't happen.

So our regional protagonists have choices. They can stop having sex with people who are likely to be infected. For Yuni, that's sex workers. For Simon, if he's in a country where HIV is already high, it is just about anybody, unless he knows they've very recently tested HIV negative, and haven't had sex with anyone else since the test, or in the few weeks before it either. Of his potential pool of girlfriends, Caspar should screen out women who shoot up drugs. He might want to avoid women whose last boyfriend was a junkie, too. But otherwise he doesn't have to worry too much about cutting down his one-at-a-time string of partners.

The second approach is to stay away from other people's body fluids. Simon, Caspar and Yuni can have sex with whomever they like as long as they use condoms with the people who are most likely to be infected with HIV. Logically, a guy like Simon who has a wide network would do well to use condoms with all his partners, and Yuni should use them with anyone he pays (even the 'girlfriend' who gives him a discount). Caspar, the serial monogamist, can probably get away without using condoms at all. But these things are circular. Simon needs to use condoms because HIV prevalence is high, and HIV prevalence is high in part because people like Simon have so rarely used condoms. Condoms are for casual sex partners, Simon believes, and none of his partners are casual. Caspar doesn't need to use condoms because there is not much HIV around, but that is in part because people like Caspar often use condoms. Condom use has risen fairly dramatically in many African countries even since I was living in Kenya a decade ago. But in most parts of Africa, men are still far less likely to put a condom on when they have sex with someone they're not married to than they are in Western countries.[28]

The third way Simon, Caspar and Yuni can try to stay HIV-free is to try and keep the virus out by making sure that sex is well lubricated and that they don't have any of the open doors that might be provided by untreated sexual infections or an uncircumcised penis. They can also hope that their infected partner is on anti-HIV treatment, and that they don't have any other STIs. Both of those things would cut down the amount of virus in their partner's body fluids, so they'd be less likely to pass on HIV. These approaches are useful at the population level, but individuals have to be mad to bet on 'well, my partner's infected but I won't get it because I'm circumcised and they don't have herpes'.

I was being unkind when I made fun of Mead for leaving the sex out of his economist's equation. There is more HIV in East and Southern Africa than anywhere else because both men and women are more likely to have several sex partners on the go at once than in other parts of the continent or the world, and because there are lots of untreated STIs and not much circumcision. Some of those factors are in turn related to some of the things in Mead's eight-part equation. The virus has spread fastest in countries where men are sucked into jobs in mines and other industries that take them away from their families. Jobs that come with cash, danger and the comfort of nearby sex workers, and that allow them to go home once a year to visit their wives. HIV has spread fastest in countries where some men have a lot of money and many women have none. These things affect how people weave their sexual networks, and changing them may well influence the 'who has sex with whom' part of the HIV transmission equation.

But the things in Mead's equation really shouldn't have much influence on whether men choose to use condoms.

Frustrated hopes

Most Brits aged over thirty-five probably remember the AIDS campaign that assaulted the nation throughout 1987. 'Don't Die of Ignorance!' we were commanded. And we were given plenty of information to help us avoid that fate. The logic at that time ran like this: AIDS is transmitted by unprotected sex and needle sharing. It is a painful way to die an early death. No one wants a painful, early death. If people know they can avoid it by using condoms and clean needles, they'll use them, as long as condoms and clean needles are easily available.

In lots of places, the logic held. In rich countries, men who had sex with one another started using condoms as a matter of course almost as soon as AIDS began its hatchet job in gay communities in the mid-1980s. In Australia, New York City, southern China, drug injectors stopped sharing needles. In commercial sex there were laggards – Indonesia was one – but in Thailand, Cambodia, Vietnam, the Philippines and large parts of China, as well as most of the industrial world, over 80 per cent of guys now use condoms when they buy sex.

In large swathes of Africa, the 'knowledge + services = behaviour change' logic didn't hold. HIV prevention programmes were mushrooming all over sub-Saharan Africa. Increasingly, the surveys showed that people knew about HIV and they knew how to prevent it. NGOs were trumpeting their condom promotion efforts. A lot of those projects focused on sex workers. But by the late 1990s, only 29 per cent of men who bought sex in Malawi used condoms. In Chad it was 24 per cent. Cameroon was more impressive at 46 per cent but only Zimbabwe got over the halfway mark at 78 per cent in a 1999 survey. And that only happened three years after the rate of HIV in sex workers in the capital Harare hit 86 per cent. Needless to say, condom use with other types of partners was much lower.[29]

This was a source of great frustration to the international public health establishment, which hadn't had a major success since wiping out smallpox in the mid-1970s. Many saw AIDS as a way to put public health back on the map, but here we were failing. Why would men in other parts of the world start using condoms in risky sex but not in the very countries where the risk was highest? It would be unthinkable to suggest that African men cared less about their lives, their families or the women they slept with than men anywhere else. Researchers bent over backwards to come up with some other explanation. And so we came back to the 'HIV is a development problem' thing. It is a convenient smokescreen to cast over inconvenient facts. And the facts are that in the worst-affected countries of Africa, most men and women at some point have unprotected sex with people to whom they are not married.

There's no doubt that HIV has *become* a development problem in Africa, because we did such a rotten job with prevention. But I don't think that HIV prevention in Africa failed initially because of poverty and gender inequity. I think it failed because most countries didn't try very hard. As late as 1999, when 23.3 million Africans were estimated to be HIV-infected, foreign agencies provided Africa with 500 million condoms. That's just over three condoms per year for each man aged 15–49. If you add in the condoms that African governments paid for, it takes it up to a grand total of 4.6 condoms per man, enough to have protected sex once every three months.[30] At the start of the epidemic, prevention programmes put the onus for condom use on prostitutes rather than on their clients, even though women don't have a penis on which to put a condom. It didn't work. What's more, some people thought that urging men to use condoms with prostitutes stigmatized both condoms and prostitutes. Prevention targeting the sex industry was sidelined even though commercial sex accounted for a high proportion of new infections.

In the very few places that *did* try hard, like Uganda and Senegal, HIV prevention didn't fail. In fact, it succeeded quite dramatically.

The countries started at different times – Senegal got prevention going when HIV prevalence was still low, and kept it that way. At the start of 2006, fewer than 1 per cent of Senegal's adults were infected. In Uganda, on the other hand, HIV had raced through the population before anything was done to stop it. By the late 1980s, when prevention campaigns got going in earnest, a quarter of pregnant women in the capital Kampala were infected, and Noerine's husband Chris and hundreds of thousands of others were dead or dying. Nationally, HIV prevalence peaked at around 15 per cent in 1991. By the start of 2006 it was down to 7 per cent.

What did those countries do? First and foremost, they dropped the 'our people don't sleep around' hypocrisy and told it like it is. They didn't pussyfoot around trying to mainstream HIV into gender-aware poverty reduction strategies. They wallpapered the country with posters telling people exactly how AIDS was spread, and exactly how it could be prevented. The state presidents talked about the disease, they talked about old men having sex with young girls, they talked about people's penchant for having several partners at once. That allowed everyone else in the country to start talking, too. In Uganda, 'zero grazing' became something of a mantra as the Simons of the world were encouraged to stop roaming around in search of new sex partners, and even to jettison extras they had already picked up. The presidents urged people to use condoms. Their rhetoric was eventually backed with massive distribution of well-packaged, cleverly marketed condoms. Senegal focused particular attention on commercial sex, which threatened to become a firelighter for HIV in West Africa. Prostitution was regulated in Senegal as it was in many former French colonies (but few British ones). So it was relatively easy for Senegal to make sure that working girls had easy access to screening and treatment for sexually transmitted infections.* Leaders in both countries made sure, too, that cabinet ministers and religious bigwigs either toed the line or kept their mouths shut, and they positively encouraged groups outside government to bring whatever they could to the prevention party. Only once they had covered the 'exposure' bases did they start worrying about other things, the things that are now so fashionable in AIDS prevention. Micro-credit schemes to give women a bigger say in family finances, keeping them from sleeping with people for money or favours, and allowing them to escape sex bondage to a philandering husband. 'Life skills' programmes to 'empower' young people, and especially girls, to control their own destinies. Legal measures that give a woman a right to inherit her husband's property, and free her from the obligation to be inherited as a wife by her brother-in-law.

There has been endless debate about which bits of the prevention efforts were most effective, especially in Uganda. 'Grazing' for extramarital partners didn't drop to zero, but both men and women whittled down the number of people they slept with. Many young people waited longer before they started having sex. Condom use rose. Was it the prevention campaigns that prodded people into safer behaviour, or the fact that people just got fed up with spending their time at funerals? Probably both.[31] Because of the public buzz around AIDS, they were certainly in no doubt what caused the funerals, or how they could avoid being next in line for a burial plot. In rural Uganda, new

* It is uncertain whether women in Senegal have enough extramarital sex to sustain an HIV epidemic, but we never got to find out. By limiting HIV in the sex trade early on and preventing a critical mass of infection among clients, Senegal eliminated the possibility of any wider spread of the virus even if the conditions for it existed.

HIV infections fell consistently since they were first measured in 1990 until 1998, when they started to level off. Worryingly, the proportion of people living with HIV seems to be on the rise again even though very few people have access to antiretrovirals as yet. There seems to be a return to 'grazing' in the last couple of years, though condom use with casual partners is steady at around 50 per cent. Researchers who have been following thousands of people in the Masaka district for eighteen years believe that risk may now be on the rise because young people who have never known an AIDS-free world simply accept the disease as an inevitable part of the landscape rather than something that they can control through their behaviour.[32] It doesn't help that President Museveni, perhaps pandering to a surge of born-again Christian conservatism in Uganda, is burying his earlier tell-it-like-it-is pragmatism. But whatever the current situation, there can be absolutely no doubt that throughout the 1990s, Senegal and Uganda made serious efforts to battle the HIV epidemic head on.

The experience of these two lonely African countries spawned a messianic belief in 'Leadership'. The heads of the United Nations, the World Bank and UNAIDS bang on and on about it. If African leaders will stand up and talk about sex and condoms the way Uganda's Yoweri Museveni did, all would be well, they imply. They are right. The problem is, much of the 'leadership' we've seen on AIDS in Africa is leading people in the wrong direction.* The most unforgivable example comes

* There's anecdotal evidence that Zimbabweans are cutting down their sex partners in response to 'Leadership' too. President Robert Mugabe's catastrophic economic policies have led hundreds of thousands of Zimbabweans into a poverty so absolute that men can no longer afford mistresses and girlfriends. See www.washingtonpost.com/wpyn/content/article/2007/07/12/AR2007071202369.

from the continent's wealthiest and best-educated country, South Africa. President Thabo Mbeki questions whether HIV causes AIDS. He has fiercely defended his scientifically bankrupt health minister, Manto Tshabalala-Msimang, who has famously advised people to eat beetroot to stave off AIDS. When she was on sick leave her deputy, Noziwe Madlala-Routledge, actually pushed through a sensible national programme to test for HIV and treat it with modern drugs. In August 2007, Mbeki sacked Madlala-Routledge, apparently because she went to an AIDS conference without his permission.[33] Ms Beetroot, on the other hand, remained in office to deal with the health needs of the 30 per cent of pregnant South Africans infected with HIV, the men who infected them and the infants who may be next in line.

You don't need to dream up complex economic formulae to explain why HIV prevention in Africa is failing. Look at the leadership. HIV would fall if fewer people had several sex partners at once, and if there were less intergenerational sex. But in Swaziland, which tops the world charts in terms of HIV prevalence, the thirty-eight-year-old king married his thirteenth bride in 2007, after she became pregnant. His new wife is younger than his eldest daughter. Look at the leadership on condoms. Men certainly won't use them to prevent AIDS if they think they don't work, or believe condoms cause AIDS. South Africa's leader conceded that condoms stop HIV, but since he hasn't conceded that HIV causes AIDS, condoms are worthless against AIDS. Kenya's religious leaders have burned condoms in public. In Mozambique and Zimbabwe, opinion leaders have spoken on the radio about how white people were pushing condoms laced with AIDS to wipe out Africans. When bishops, presidents and the media poke holes in condoms, they become ineffective.

Sensitive souls

By the time Ronald Reagan forced the word 'AIDS' out of his mouth in late 1985, 12,689 Americans had died of the disease. By the time UN Secretary General Kofi Annan addressed African leaders on the subject in 2001, over 28 million sub-Saharan Africans were living with HIV, and over 18 million had already died. That's 46 million people infected with a fatal disease that you get mainly by having sex with someone who is not your sole, lifelong, faithful partner. And still many African leaders were in denial. For a moment, it looked as if the Ghanaian Secretary General was going to try to shake his peers out of their fantasy world of abstinence and monogamy. 'The epidemic can be stopped, if people are not afraid to talk about it,' he said. 'You must take the lead in breaking the wall of silence and embarrassment that still surrounds this issue in too many African societies.' But he did not himself wield the sledgehammer. In his fifteen-minute speech he didn't mention sex once.[34]

He's not the only one who finds it hard to speak about sex.

Force-of-Nature Noerine was one of the very first people to punch a hole in the 'wall of silence'. In 1987 her husband Chris died of AIDS. The disease, known in Uganda as 'Slim', was beginning to creep into people's loves, lives and consciousness. But nobody had had the courage to stand up and say, 'Someone I love has died of AIDS. AIDS is in my family. AIDS is in my life.' Noerine did it. And she has had to do so over and over again in the last two decades, as one member of her extended family after another has been mown down by HIV. She gathered up her strength and spoke in public about HIV, about sex, about the need to use condoms. She started The AIDS Support Organization (TASO), Africa's first such support group. She acted as a counsellor to thousands of people who were infected, dying, or sweeping up the fragments of a shattered family. As an AIDS-affected African female, she enjoyed a hallowed status on the politically correct international AIDS circuit. She used it to preach the message of safe sex to policy-makers in Uganda and internationally. And yet when it came to talking to her own daughters about sex, she simply couldn't do it. 'I left a box of condoms in their bedroom,' Noerine giggles. 'I just hoped they'd figure it out.'

Where did this wall of silence come from?

What little we know about pre-colonial sexual behaviour in Africa comes from a few white men – explorer Richard Burton and his ilk.[35] They portray cultures that embrace sex with the joyful enthusiasm it deserves. Much was made of the size and adornment of men's genitals. The whole was presented as rather – how shall I put it? – *glauque*. That wonderful French word again, seedy and unpleasant and definitely to be disapproved of but somehow rather thrilling, too. The fact that the Victorian accounts of sex in Africa were *glauque*, voyeuristic, lacking in statistical significance, did not necessarily make them untrue.*

Back in London, the accounts were seen as confirming that the civilizations of Africa were anything but civilized, definitely not up to the moral standards of good, upstanding Christians. The colonists soldiered on with their 'civilizing' mission. In time, they managed to stamp Christianity on much of Africa. The rhetoric of chastity and monogamy was adopted wholesale. But as we have seen, the behaviours were not.

* As far as genitalia go, today's condom companies appear to agree with the Victorians. Condoms are made in three main widths: 49mm dominates the Asian market, 52mm is the norm in Western markets, while in Africa up to 54mm is common. For more details about international standards and specifications, see www.fhi.org/en/RH/Pubs/booksReports/latexcondom/standspec tests.htm.

I believe that the disconnect explains to some extent why African leaders have been so reluctant to talk about the realities of sex in their communities. Calling your voters hypocrites does not play well at the polls. And voters do not seem to be clamouring for greater honesty. When I was in Kenya, I remember parents demonstrating outside a technical college that had introduced sex education lessons for pupils in their late teens and early twenties. They burned condoms, and waved placards equating premarital sex with beastly behaviour: 'Our children are not doggy-dogs!' they protested. Sex education was dropped from the curriculum.

A more recent example of citizens burying their heads in the sand comes from Swaziland, where 43 per cent of adults are infected with HIV. I've been in the AIDS business for more than a decade and I still can't get my head around numbers like that. I can't get my head around this, either: After years of bland 'Be aware: AIDS leads to death' type campaigns, Swaziland's AIDS council in 2006 endorsed a new approach. Ignoring the king's thirteen wives, they ran a campaign that drew a clear connection between multiple partnerships and AIDS. 'OK honey, you have a good time at the conference in Durban,' says a cartoon businessman, before texting his secretary: 'The wife has gone to Durban. Come over tonight and roll in the sack with me.' His secretary texts back to reject him: 'I am no longer your roll-over.' Pretty tame. But within weeks, the Swaziland National Network of People Living with HIV and AIDS had organized a march on parliament and got the campaign, called 'Secret Lover', taken off the air. According to a press release issued by the International Community of Women Living with HIV/AIDS: 'A new government-sponsored HIV prevention campaign in Swaziland uses insulting language to target HIV-positive women and suggests that they are the cause of the spread of HIV.'

'The reality is that most Swazi women "get HIV in their own bedrooms from their husbands",' said a member of the International Community of Women Living with HIV/AIDS in their press release. But epidemiologically, it is hard to see how that could be the case. Some husbands will have been infected by sex workers. But a goodly proportion will have been infected by someone's current or future wife, including their own. HIV will only ever reach those levels of infection if a majority of the population has unprotected sex with more than one partner. Apparently, though, we're not allowed to say that men get infected by having unprotected sex with HIV positive women, each of whom herself became infected by having unprotected sex with an HIV positive man. Here's something else we're certainly not allowed to say: with that kind of guidance from the very people who are most affected, it's no wonder that more than two in five adults in Swaziland are infected with HIV.

Kofi Annan's 'wall of silence' certainly proved difficult for African leaders and communities to scale. But why did the international community aid and abet them by coming up with the 'AIDS is spread by poverty' rationale, instead of punching a hole through taboos about sex? Perhaps because well-meaning save-the-world types face a far greater taboo: race.

The fear of appearing racist has affected how we analyse data and how we present the results. It extends even to such fearless organs as *The Economist*, a weekly newspaper that happily argues in favour of legalizing drugs and allowing gays to marry. I have always loved *The Economist*'s tell-it-like-it-is style. So when they asked me to write a piece about AIDS in Africa ahead of the International AIDS Conference in Durban in 2000, I told it like it was. I presented data showing that in many countries in Africa people start having sex younger, have more partners at any one time and have sex more frequently than people in other parts of the world, and they are also less likely to use condoms. The story never ran. The deputy foreign editor told me that the Africa editor at the time, a white Englishman, had objected that it was racist to say that Africans had more sex than Asians, Europeans or Americans.[36]

Here's my question to white Englishmen: how is it a *bad* thing to say that people have a lot of sex?

Most Western countries have thrown off the stuffy Victorian morality which seems still to cast its shadow over Africa – the former British colonies in particular. To most of us, sex is not something dirty or shameful or wicked. It is something fun. But if you add HIV to the mix, the shame factor seems to raise its head again. To understand why, we have to go back to the beginnings of the epidemic in Western countries.

HIV burst into the world's consciousness through gay men in rich countries, especially the United States. In the early 1980s, a number of noisy and self-righteous Bible bashers greeted AIDS with some delight as a sign of God's retribution for the immoral and unnatural act of sodomy. This is nonsense of course. But if you take morality and retribution out of the equation, the Bible bashers had a point. The more people you have sex with, the more likely you are to have sex with someone who is HIV-infected. If you have sex in ways that do not follow basic human sexual design (which includes a lubricated vagina), you will increase the chance of small tears and abrasions which allow the virus to pass from one body to another. If you switch between receptive anal sex, in which you're very likely to get infected, and insertive anal sex, in which you're very likely to pass on infection, you're going to speed up HIV transmission. HIV is not divine retribution for unprotected anal sex with lots of other people. It is simply a *consequence* of unprotected anal sex with lots of other people, in the same way that lung cancer is a consequence of smoking, and obesity is a consequence of eating fast food, drinking super-sized Cokes, and getting in your truck to drive the 800 yards to church instead of walking.

It didn't help that HIV is also a consequence of sticking needles carrying HIV-infected blood into yourself, and that the people most likely to do that were drug injectors. In the early years of the epidemic in the West, having AIDS was like having 'Wicked Person' branded across your forehead.

The gay lobby, full of well-educated and articulate men with good access to the media, fought back against the deliberate conflation of biomechanical causation and morality. They battled vociferously for the rights of those infected, the right to jobs and services, the right to be treated with dignity, the right to be treated at all. They battled, too, for the right to anonymity and confidentiality. Their success in the United States was mixed. But the philosophy underpinning their battles was adopted wholeheartedly by the international AIDS mafia, led at the time by the redoubtable Jonathan Mann. Mann, the first head of the WHO's Global Programme on AIDS, exported the 'rights-based approach' to HIV to Africa with grim determination.

Public health is inherently a somewhat fascist discipline. It accepts that we must sometimes violate the rights of a few to protect the health of the many. Look at SARS. When that crawled into view we slapped restrictions on people's movements, forced people to be tested and registered them if they were infected, even shut down whole cities. Since the nineteenth century, many countries have registered people with syphilis, and have asked those people about their sex partners. Those partners have been tracked down and, latterly, offered tests and treatment. No one complained much, even though syphilis was once every bit as stigmatized as HIV is now. But when authorities in the United States, Sweden and Bavaria tried to use the same measures against HIV early in the epidemic, everyone complained a lot, and many key measures were quashed.[37] AIDS focused people's attention on the rights of the people living with an infectious disease. This was a new departure in public health, and a welcome change in many ways. But the fear of violating people's perceived rights overrode many otherwise routine principles of public health.

By the time we began trying to address HIV in Africa, mandatory HIV testing and contact tracing were philosophically out of the question. Completely confidential and strictly voluntary HIV testing were the order of the day.

Making a big deal out of confidentiality sends mixed messages. 'HIV is not a retribution, it's not a scourge, you don't have to feel guilty about being infected, you're no different from anyone else' on the one hand. But on the other: 'Don't worry about being infected with this dreadful disease. We promise not to tell anyone, not even your husband, because if we did tell him, who knows what might happen to you . . .' This was not unreasonable in the circumstances. Swarms of people have lost their jobs because they were HIV-infected. Tales of women getting thrown out of their homes by the very men who infected them are rife. Parents abandon their own sons. In December 1998 a South African woman called Gugu Dlamini was beaten to death in KwaZulu Natal province after she revealed that she, like one in five adults in South Africa at the time, was HIV-infected.

But stigma feeds on itself. Think about countries like Swaziland, where more than two in every five people are infected with HIV. Whatever the inherited Christian rhetoric says, multiple partnerships must be the norm – you can't reach those levels of heterosexual infection any other way. And still we brand the behaviour 'wicked', and treat AIDS as a badge of dishonour? It is insane. If everyone were open about their behaviour and their infection, stigma would dissolve simply because there are not enough unaffected people left to do the stigmatizing.

Export-led growth

Here's the difference between Africa and the rest of the world. In much of East and Southern Africa, HIV is spread by something most people do. In the rest of the world, it is not.

In East and Southern Africa people have sex in nets not strings. There are plenty of other sexually transmitted infections to push HIV through those networks, and not much circumcision or condom use to put a brake on it. Very quickly, HIV really was everyone's problem. And that made it a development problem, increasing poverty, swamping an already rickety health system, and hollowing out schools and businesses.

Countries such as Senegal and Uganda, countries that have talked about extramarital sex instead of just pretending it doesn't happen, have either staved off an epidemic or reversed it. And they've done it without governments doing any of the peripheral development stuff. A few other countries are trying to follow suit. But for whole generations in many countries, the change has come too late. For the same amount of sleeping around, you now have a greater risk of getting infected with HIV if you use a condom every single time you have sex in Swaziland than you do if you never use a condom at all in China. The likelihood of a condom bursting is tiny, but where close to one in two of your potential lovers is infected with HIV, as in Swaziland, the combined probability of burst condom plus infected partner is still higher than the likelihood of chancing upon an HIV positive heterosexual partner in China. In fact, it is higher than chancing upon an HIV positive prostitute in China.[38]

The 'HIV is a development problem' chorus has become a self-fulfilling prophecy in Africa. It does seem likely that patterns of sexual behaviour will change faster where there's more equality between men and women, when there's more disposable income in households, where the economy brings families together instead of splitting them apart. But changing those things is a very long-term goal. In high prevalence countries, if people don't become more discerning about their sex partners and governments don't become more aggressive about ensuring cheap and easy access to condoms, STI treatment, circumcision and antiretrovirals, there won't be a long term.

In the rest of the world, the idea that we have to change the fabric of society before we can address AIDS is not remotely plausible, because HIV is not something that permeates the whole of society. 'People get AIDS by doing things most people do not do, and of which most people do not approve,' said Everett Koop, US Surgeon General at the time the AIDS epidemic emerged in America. He was talking about the US in 1988, but he could equally have been talking about 90 per cent of the world two decades later.[39] Now that we've made sure blood transfusions are safe, most new HIV infections in Asia, Europe, the Americas, Australasia, the Middle East, North Africa and even parts of West Africa are the result of drug injection and anal sex between men, as well as between people who buy and sell sex. It has been like that for twenty years, and it's not going to change.

We could damp down HIV across vast swathes of the globe just by being honest about who is infected and why, and by giving them the information and services they need to interrupt transmission. And we could probably do it for less money than we already have for prevention. But instead of getting on with it, we're busily importing the 'HIV is a development problem and everyone is at risk' smokescreen from Africa.

Policy-makers in Asia are delighted because it means they can get their hands on AIDS money without having to think about drug injection, male–male sex or prostitution. Understandable. If you were a politician which would you rather deal with – a development problem or a heroin problem?

Activist groups have bought into the rhetoric too, because they don't want to add to the stigma with which sex workers, drug injectors and gay guys already grapple. Better 'Everyone's at risk' than 'Only wicked people are at risk'. And a lot of public health types have gone along with this, even though we know it's not true. Our greatest fear is that if we stand up and say, '*Not* everyone is at risk! Only the people you consider the dregs of society are at risk!', people will say 'Great! We were wondering how to get rid of the dregs!' Then HIV prevention will drop off the map completely.

But who are these 'dregs'? If not us, then probably someone close to us. They are our brothers and our lovers, our husbands and our friends. They are Nana, who is supporting her three kids, and Bhimanto, who is a respected foreign correspondent, and Lenny, who ran a first-class restaurant and now runs a first-class NGO. They are the pilot who is flying your plane, and the boss who is paying your salary, and the teacher who is schooling your kids. As we found in our 'Best Practice' estimates, there are 5 million 'dregs' just in Indonesia, and another 41 million in China. India's estimates are less rigorous, but they seem to point to something like 38 million people buying or selling sex, injecting drugs, or having anal sex with several partners.

Very few of these millions are school children or pregnant women or migrant workers. But here we are spending money on life skills programmes in schools, HIV testing for pregnant women and posters in factories. None of these are bad things. But we know perfectly well that they do virtually nothing to interrupt HIV transmission in most of the world. Touchy-feely programmes for the general population do not make drug injection disappear. They do not make sex between men evaporate. They might whittle away at commercial sex, but they won't eliminate it. It wasn't treating HIV like a development problem that put the lid on the epidemic in Australia, Great Britain, Canada and Japan. It was treating it like an infectious disease which could be prevented by providing services that were wanted and needed by people who have anal sex with lots of partners, people who buy and sell sex and people who inject drugs.

We've got ourselves into a giant poker game. In poker, you have to throw away money on hands you know are not going to win, so that you can lull people into letting you get away with the really important bets. Many public health pragmatists believe that it doesn't really matter if we waste money on programmes for schoolgirls and housewives and Boy Scouts, as long as we can do what we need to for junkies and gay guys and the people who buy and sell sex. When Margaret Thatcher's government put 23 million leaflets about AIDS through Britain's letter-boxes in 1987, the HIV prevention mafia knew they weren't preventing many infections. But the hoopla created a lot of public concern, and that in turn created elbow room to do the things that were really needed, the down-and-dirty work with people who were most at risk.

In many African countries HIV has slashed life expectancy back to levels that we haven't seen for half a century or more. Here, it is absolutely essential to spend money testing pregnant women for HIV, teaching village health workers how to spot and treat sexually transmitted infections, teaching schoolchildren how to talk about sex and condoms. In the rest of the world, these things won't make the tiniest dent in HIV, and we know it. We comfort ourselves by saying they might have other benefits, that at worst they are not actually harmful. If it is the price we have to pay to hide the realities of HIV from the electorate, so that we can quietly get on with doing the things that really need doing for the people 'most people disapprove of', then that's fine with me.

But here's the problem. In a lot of countries, we're not doing the touchy-feely everyone-is-at-risk programmes to distract attention or buy public support so that we can get on with the programmes that will actually reduce the spread of HIV. We're doing these things *instead* of the things that could slash new cases of HIV. We created the 'HIV is a development problem' mantra because neither African leaders nor the international public health establishment wanted to talk about sex in Africa. It became the cement in Kofi Annan's wall of silence. We've exported that cement to countries where leaders don't want to talk about drug injection, or commercial sex, or anal sex. Now we're building walls of silence around drug injectors, gay men and the sex trade, and that makes it hard to do what's needed to shut down HIV.

Back in Asia, I was discovering that even behind those walls, even where we do have prevention programmes for people most at risk, a herd of ideological sacred cows can trample HIV prevention efforts.

5

Sacred Cows

I was once described in a book about AIDS in Asia as 'AIDS activist Elizabeth Pisani'.[1] I found this mildly amusing – I don't consider myself an activist, I'm just a nerd who is trying to persuade people to look at the facts and do something sensible about them. But it got me thinking. You never hear of a flu activist, or a syphilis activist, or even a cancer activist. But 'AIDS activist' trips off the tongue nicely. The HIV prevention industry is rooted in the gay bars of New York and the bathhouses of San Francisco. It was fertilized by self-reliance in the face of discrimination against gay men, and its fruit was a culture of activism that has been plucked and exported around the world. As we've seen, that has had a profound effect on the way we've addressed the virus in other parts of the world.

We've inherited a lot of sacred cows from the admirable godfathers of activism, and we have rarely dared to question whether we really want or need them. One, as we've just seen, is that HIV testing exposes people to stigma, so routine testing is dubious and compulsory testing is evil. Others relate to who does what best in the world of HIV prevention and care. The holiest cow in the herd is antiretroviral treatment: an unqualified Good Thing, because it acts as effective HIV prevention as well as keeping people alive.

I accepted a lot of the conventional wisdoms when I started in the AIDS business. But in the murky world of brothels, shooting galleries and gay bars, the hard edges of certainty and conviction begin to blur. As the data began to roll in, I asked questions of it. A few of the sacred cows remained holy. But many were sacrificed on the untidy altar of Reality.

Less is more

Let's start with the holiest of the activists' cows. Without doubt the greatest impact that AIDS activists have had is on access to treatment. The gay lobby piled on the pressure for a cure from the early 1980s. The pharmaceutical industry responded with a massive research effort. They have failed to produce a cure, but they came up with pretty effective treatments. The drugs work in several different ways to stop the virus replicating. Most of them mess with various proteins that HIV needs to make copies of itself. The newest types of drugs put glue on the key that HIV uses to unlock and invade human cells.

Because HIV reproduces so swiftly and in huge numbers, it can quickly generate mutations that can help it beat a single drug. To stay ahead of the virus, people have to take two or three different drugs at once. In the early days, each drug cost thousands of dollars per person per year – only people in rich countries with good health insurance schemes or very generous governments could hope to get treated. Then the lobby of HIV-infected people in the developing world weighed in, shaking Big Pharma to its very foundations. Their vocal demands have brought the price of HIV drugs crashing down; services are now being provided to millions of people who just a couple of years ago could only dream of such assistance.

Activists quite rightly hail this as a great victory; they continue to press loudly for treatment to be expanded further. They are less loud about the dangerous edge that goes with increased treatment. Ask most people in the AIDS industry about the relationship between antiretrovirals and prevention, and here's what they'll tell you: HIV is more easily transmitted when there's lots of virus in the blood. Antiretroviral medicines push the amount of virus down. So obviously, the viral load in someone on ARVs will be lower than that of someone who's not on treatment. And that means they'll be less likely to pass on HIV. There you go. Treatment *is* prevention.[2]

Not so fast. What is the viral load of someone with AIDS who is not taking AIDS medicine? It shoots up for a while, maybe a year, possibly two. And then it drops to zero, because the person dies. The person on therapy can carry on for years and years – we don't know how long yet, because the drugs are too new. On the whole they'll be in pretty good shape. Good enough shape to go out dancing, to sell sex, to party with friends and shoot up drugs, to pick up that cute boy who works behind the bar at Prego. They'll probably have the odd spike in viral load now and then even when they are on medication, if they get the clap or some other infection. Someone whose viral load is low with the occasional spike and who is out getting laid is certainly more likely to transmit HIV than someone who is dead.

Once you start taking AIDS drugs you don't stop until you die, and that could be very many years from now. So drugs must be provided for everyone who is already taking them, *plus* the people who develop AIDS this year and next, *plus*, eventually, all the people who got infected this year because we were working so hard to increase access to drugs that we forgot about prevention. The number of patients in need of medication quickly adds up.

The more effective our prevention programmes are, the less treatment we will need. But the more effective our treatment programmes are, the longer people live with their infections, the healthier and more sexually active they are, the *more* prevention we will need.

There's something else going on, too. Treatment makes HIV much, much less scary, because it makes it less fatal. There are fewer cadaverous people around, fewer funerals to go to. With treatment, people who were at death's door leap up and march back to the office and the nightclub. With treatment, you can ask 'how's your brother?' questions you used to be afraid to ask in case the person you were asking about was dead. All of this is wonderful. But . . . (With AIDS, there always seems to be a but . . .)

As people get less scared of AIDS, they get sloppier about prevention. One of the swiftest prevention success stories was among gay men in the industrialized world. Gays in San Francisco were the Formula One drivers of condom adoption – condoms in anal sex zoomed from zero to around 70 per cent by 1985, and condom use hovered around that level until 1994. Then along came treatment. AIDS got less scary, people got bored of being 'good', there were guys in the bars too young to remember the days when every weekend was a wake. Rectal gonorrhoea, which had halved between 1989 and 1994, did an abrupt U-turn when HIV treatment became the norm, doubling back to its earlier levels by 1998. And then it kept going up, doubling again by 2005. Same story for syphilis. There really wasn't any syphilis among gay guys in San Francisco in the late 1990s. In 2004 around 250 cases of early syphilis infection were reported in the city. Why all this disease? Because condom use dipped from its peak levels. By 2001, half of all HIV positive men had tossed condoms aside. Of course that wouldn't really matter if they were having sex with people who were also infected. But in 2001, close to a third of men who had tested positive didn't bother to use condoms with a guy who was negative, or whose HIV status they didn't know. This was up from a fifth in 1998 when the question was first asked. Despite all those people on treatment, with their lower viral loads, new HIV infections among men seeking testing more than tripled, from 1.3 per cent in 1997 to 4.7 percent two years later.[3] Similar rises in risk and infection rates among gay men have been recorded in London, Amsterdam and Vancouver.[4] People get sloppy about prevention partly because they're thinking: 'Well, if he's got AIDS, he's probably on treatment, so he won't be that infectious.' But remember that people are *most* infectious when they first get infected, before they even test positive for the virus, let alone start taking antiretrovirals. If condom use falls, newly infected people have a better chance than before of passing on HIV, regardless of whether they eventually get treated.

The minute you open your mouth with a 'yes, but' about antiretroviral treatment, you are branded a monster. 'So what are you saying, that we should just let people die?' Of course not. But I don't think it is fair to demand that people be honest about sex and drugs, and then refuse to be honest about what will happen as we expand access to treatment. More treatment means more people with HIV, potentially taking more risk and exposing more other people to the virus. We need to make sure that we have the money and the staff to bump up prevention services as more people get treated. And we need to see treatment as an opportunity to draw people into prevention programmes, rather than let people think: 'I'm on antiretrovirals, I'm alright, Jack.'

At the moment, treatment seems to be elbowing prevention off the global stage.[5] Treatment is popular – voters smile on compassionate politicians who give out life-saving drugs, while frowning on politicians who give out condoms or clean needles that would make those drugs unnecessary. The World Health Organization rode into battle with AIDS in the early 2000s under the banner of 'Three by five' – the idea was to get 3 million people in the developing world on ARVs by 2005. We missed that target by a very wide margin, but it did serve to divert attention away from prevention. In Thailand, held up as a model of dedication to HIV prevention, treatment is eating up more and more of the AIDS budget. By 2005, only 13 per cent of all HIV money was spent on prevention.[6] And on the front lines, the very people who have been leading the prevention charge are getting overwhelmed with the needs of people who are sick.

'To be honest, it is really depressing to be going around talking about prevention and prevention and prevention when we know perfectly well that maybe a third of the people we're talking to are already infected,' said my friend Lenny, who runs a programme for fellow waria in Jakarta. 'Of course we have to shift our attention to helping people who are already sick.'

I'd dropped in for a gossip with Lenny on a visit back to Jakarta at the end of 2006, more than four years after we had worked together on that study of HIV in waria that made me cry. Lenny had been bustling around in the back room of her salon as we spoke. She'd had a long day. After a morning meeting with the United Nations Population Fund about condom distribution, she had slunk home to rest. When I rocked up to visit she was wearing a hideous baby-blue cheesecloth smock – pure 1972 hippie chick. The brown woollen bonnet she wore to cover her hair was doing her no favours either. But half an hour in the chrysalis of the back salon, and she emerged transformed. The floral chiffon blouse would have done justice to Christian Lacroix; it wafted over elegantly flared black trousers. High-heeled sandals had replaced her plastic flip-flops, and the bonnet was gone, revealing carefully highlighted hair. She was off to Bangkok, to argue for the inclusion of transgenders in the Asia Pacific Network of Sex Workers, 'in my cowboy English', she laughed. She was only going for four days, but her suitcase was the size of Lenin's mausoleum. 'It's important to have good outfits.'

This was not the first transformation I had seen in Lenny. The last time we had met, about eighteen months earlier, she was a patient on a male ward in a public hospital.* Her hair was covered in another ugly bonnet, she wore no make-up and her baggy yellow T-shirt managed at once to camouflage her silicone-enhanced breasts and show off her well-muscled arms. But she still looked massively out of place among the old men on respirators with whom she shared a room. Lenny's HIV infection was getting the better of her. Later, the white blood cells that would normally provide defence against disease all but evaporated, crashing from a normal level of about 800 cells per millilitre of blood to just seven, 'practically dead, really'. But she had recently been put on a new combination of drugs to control the virus. Now she looked like a million dollars.

This snatched-from-death experience has a name in the industry: the Lazarus transformation. A lot of the waria in Jakarta had seen Lenny go from her graveside back to her parties, her

* One of the reasons waria have so many health problems is that they avoid going for treatment when they need it. They hate being herded into the male wards and clinics at government health services.

meetings in parliament, her conferences in Bangkok. Many of those who were getting sick wanted a bit of Lazarus for themselves, and God knows Lenny and her colleagues want to provide it. 'We were way too late with prevention, but all that outreach work we did was like sowing seeds. Now it's time to go out and harvest.' And then, in case I missed the point: 'Harvest people with AIDS, I mean.'

Testing truths

When the Western activists exported their horror of any HIV testing that was not voluntary and entirely confidential, there was no treatment available. Effective drugs began to flood into rich countries from around 1996, but they took another decade to trickle through to people who needed them in the developing world. When Lenny and I worked together on our study of waria in Jakarta in 2002, antiretrovirals were still a distant dream. When we offered HIV tests to waria as part of our research, we'd quite often get an are-you-nuts? kind of stare. An HIV test? Why? Just so that I can get stressed out? Most waria allowed us to take blood just to be polite, and to get the giant tube of lube we were offering to study participants, but only one in seven came back for her test results.

Now that we can do something for people with HIV, it is a different story. Many more people are seeking testing in developing countries, especially people who think they may be infected. That has sent reported HIV cases soaring. China, for example, said reported HIV/AIDS cases had jumped by over a third in 2005. This was seized on as a sign that the epidemic was exploding in the world's most populous nation.[7] Should we panic? No. Reported cases rose because in 2005, China went out and tested hundreds of thousands of people at high risk of contracting HIV. In other words, they were finding more HIV just because they were looking for it. We've had to abandon the 'tip of the iceberg' metaphor because the hidden part of an ice-berg never emerges, where as the hidden HIV epidemic is beginning to show. In Indonesia some now compare the HIV epidemic to a water-buffalo slowly pulling itself out of the mud to reveal its true size.

The sacred cow of 'only voluntary confidential testing' is beginning to totter. I saw this clearly in the data from China. Always a law unto itself, China started steam-rollering people into finding out if they had HIV in 2004. Why? It goes back to a horror story from the mid-1990s, when penniless peasants were selling their blood to fly-by-night medical supply firms. The blood was mixed up together, the money-spinning plasma drawn off, and the red blood cells pumped back into the peasants, so that they could give more blood again sooner. Unfortunately someone threw their HIV-infected blood into the mix, and soon the virus had stormed through whole villages. Local authorities did what Chinese mandarins do when some-thing goes horribly wrong – they covered the whole thing up. Then in 2001, along came *New York Times* reporter Elizabeth Rosenthal. A delightful doctor-turned-journalist, Elizabeth was relentless in her pursuit of the infected blood story in China; she blew the cover-up to smithereens. There were plenty of people in China's Ministry of Health who wanted to take on the AIDS problem, but they were walled in by politics. Elizabeth did as much as any other individual to smash down that wall and set the technocrats free to do their work.[8] By 2004, China had announced the 'Four Frees and One Care' policy. This promised everyone free voluntary counselling and HIV testing, and anyone with HIV free antiretroviral drugs, free prevention of mother-to-child transmission and free schooling for their children. Care centres were set up nationwide. Still people were reluctant to come forward for testing. What would happen if they were infected? Would they end up in jail? Would they lose their jobs?

With so few takers for the Frees and Cares, Beijing decided to go out and find people who needed the services, hence the mass screenings. In Henan province, the dirty-blood basin of China, over 258,000 former plasma donors were identified and tested for HIV in 2004, whether they liked it or not. Over 23,000 of them turned out to be infected. Did they lose their jobs? No. Did they go to jail? No. Miraculously, they got pretty much what the government had promised.* Suddenly, plasma donors who had 'escaped' the mass testing started coming out of the woodwork asking for HIV tests. Nearly 7,000 of them, in fact. HIV prevalence was twice as high among those who asked for tests as it was in the compulsory mass testing. This suggests that people who suspected that they might be infected had hung back to see what would happen to people who tested positive.

What does this illustrate? In part, that bribery works. Where we really *do* provide people with antiretrovirals, and decent health care and schooling for their children, people will overcome their fear of stigma and come forward for testing. But in part, too, that a certain amount of routine or even involuntary testing can help break the dam of stigma, can show that it is not so bad to be tested. It can get people into services to which they wouldn't otherwise have had access. And it can help to increase demand for voluntary testing. Granted, plasma donors

* Free education is a valuable commodity in China. According to a colleague in Beijing, this has led to a secondary market in HIV positive blood – people are faking being infected so that they can send their kids to school.

are seen as 'innocent victims'. Mandatory testing may or may not prove to be less benign among drug injectors and other risk populations now being rounded up and screened in China. But we shouldn't just assume the worst.

When I first heard about the mass testing in China I had reacted with all the predictable, knee-jerk comments about a violation of ethics, about protecting people's right not to know if they were infected, and their right not to tell anyone else if they did know. But now that we can do something useful for people who are infected, testing the people most likely to be infected is beginning to make sense. Could it lead to people being outed, being stigmatized? Yes. But untreated AIDS has a way of outing people anyway, so you get the stigma *and* the avoidably early death. What should doctors do when they know that an HIV-infected person has not told their regular sex partner? If a doctor insists on informing that partner, he or she may be exposing their patient to discrimination or abuse. But not telling exposes the partner to the risk of being infected with a fatal disease, or denies them the chance to get treated if they are already infected.

Think of it this way. You're in a bar with your best friend. She's being chatted up by a guy who you happen to know has HIV – you represented him in a lawsuit that infected haemophiliacs filed some years ago. She's getting hot and heavy with this guy, but before they leave together he goes to the bathroom. You know from experience that your friend is a slob about using condoms. Now you're alone with her. Do you say anything?

This is the dilemma that doctors now face, and more and more of them are arguing that they have a responsibility to the sex or injecting partners of the person who is infected. The public health establishment is beginning to suggest that we should be less paranoid about HIV testing. New WHO guidelines urge health services to tell patients they will be tested for HIV, and then to go ahead and test unless the patient actively objects.[9]

Activists who ripened in a culture of 'confidentiality above all' are still uncomfortable with the shift, worrying that health services will provide the testing, but not the medicines and prevention services that make a test worthwhile.[10] They'd doubtless be horrified by what I see as the logical next step – requiring people who are getting free AIDS drugs to show up for prevention services, too. You could call this bribery, or you could call it bundling. The same way software companies bundle some browsers you don't really want with operating systems you really do want. Most people will use what they're given, even if they don't like it that much. The fact is, most people infected with any disease care more about treatment than prevention. We could use antiretroviral drugs as an incentive to take methadone, to attend a needle exchange programme, to go to the clap clinic for regular check-ups, to do the things that will reduce the likelihood that you'll pass your infection on. Some people will opt out, taking the treatment but not the prevention. But my guess is that a lot will accept prevention services in order to get their antiretrovirals, even if they don't like it that much.

I should make it clear that people only get het up about HIV tests which are forced on people, or which people feel obliged to go along with. Everyone is hugely in favour of truly voluntary testing, in part in the belief that if people ask for a test and get counselling before and after it, they'll be less likely to take risks in the future. The theory is that if you test positive, you'll jump at condoms to avoid passing the virus on. If you test negative, you'll jump at condoms to ensure that you stay negative. But the data show that it doesn't always work like that. In rural

Zimbabwe, women who are counselled and then test HIV positive use more condoms than women who have never been tested, which is encouraging. But those who are negative go on blithely with the same risks. HIV-infected drug injectors in Bangkok use condoms more often after their diagnosis than they did before, but go on sharing needles at the same rate. As we saw earlier, despite all the counselling that came with their diagnosis, a third of gay guys with HIV in San Francisco were at one stage having unprotected anal sex with men who may be negative. In Indonesia, female sex workers who have been counselled and know their HIV test result are more likely to use condoms than those who've never been tested, but waria, male sex workers and gay men are less likely to use protection.[11]

For some people, it seems to work like this. If you test positive: 'I'm already infected, what's the point of using a condom?' If you test negative: 'Well, I've got away with it so far, I must be doing something right. See, I knew I could pick the nice boys . . .'

At the moment, we are so focused on trying to increase the number of HIV-infected people on antiretroviral treatment that we tend to let people who test negative go away with nothing more than a handful of condoms and a little lecture about staying safe. But more voluntary counselling and HIV testing does not necessarily translate into less risky sex or drug injection. What it does do is provide an opportunity to hook people into active HIV prevention programmes that provide the services people need. We should stop squandering that opportunity.

Small is beautiful?

When the US government buried its head in the sand when AIDS first emerged, the gay community was forced to organize itself. The people handing out safe sex packs in the gay bars I used to go to in New York in the early 1980s were not public health workers. They were gay guys who responded to the battle cry of the pioneering self-help group the Gay Men's Health Crisis. By the time the government got on board, these organizations had cornered the market in HIV prevention. Public health authorities were forced to seek their advice, their help and their expertise. The snobbish medical establishment had never done this before. Doctors are used to giving advice, not asking for it, especially not from a bunch of guys in leathers and earrings. The partnership proved quite effective. But the experience bred some more sacred cows. One was a belief that NGOs could do better at HIV prevention than government. A second was that the best people to provide prevention services were 'peers', people who were themselves members of the affected groups. A third was that no HIV programme can succeed unless people infected with the virus are actively involved in planning the programme, and often in delivering services too.

When I arrived in Indonesia, I didn't have enough experience to know whether these received wisdoms were really all that wise. But I was going to have a chance to find out, because NGOs and 'peer outreach' were at the core of the HIV prevention programme FHI was managing.

Non-governmental organizations can reach places governments can't reach, the theory goes. That's often true. Small NGO-run interventions can make a huge difference to the lives of the people they touch. The problem is, they don't touch many lives. Local NGOs are like high-fashion boutiques. They sell very high-quality products – the Prada bag, the Armani frock – to a small number of people at very high prices.

I have the highest respect for NGOs that bend over backwards to serve their clients. They have blazed trails through territory that governments wouldn't even look at. They have provided services that the big funding agencies didn't want to touch. They have even provided services their own higher-ups might not want to touch. One of the better-organised sex worker clinics I worked with was run by Catholic nuns in the Indonesian outpost of Papua. As prostitutes waited for their health screenings, the more musical nuns would strum out safe sex songs on their guitars.* They have discovered what is necessary and demonstrated what is possible. But they have done it on such a small scale that it has made no difference to the epidemic.

When I arrived in Indonesia in 2001, FHI had been helping small NGOs work on HIV prevention for five years with sex workers in the capital Jakarta, the largest port city Surabaya and the northern port of Manado. In that time, condom use in commercial sex had barely budged. If we were delivering Prada quality, why didn't condom use soar, at least in the cities where we were working? Looking at the 2002 data, I could take a pretty good guess. Even in programmes that had been running for six years, only 7 per cent of prostitutes reported having talked about HIV with someone from an NGO, and among newer programmes rates were far lower. One of the most depressing graphs I ever made was one showing the impact we would have on the HIV epidemic in Indonesia if every single person we reached in 2002 cut their risk behaviour to zero overnight. Even in that pigs-might-fly scenario, we'd only reduce new HIV infections by 3 per cent, in a programme that was costing around US$8 million a year.[12]

Producing Prada quality for a handful of people is a good way to get started. But to prevent an HIV epidemic we don't need

* Around a quarter of the US-funded NGOs working on AIDS in developing countries are faith-based. Figures for other countries are not available.

perfect services for a few people. We need 'good enough' services for everyone who needs them. We need to shift from the boutique, community-driven, bend-over-backwards approach and go instead for the discount supermarket approach. Pile it high and sell it cheap.

Look around. Who provides mediocre services to vast numbers of people? Governments. But can civil servants reach sex workers, drug injectors, gay guys? And can they provide them with adequate HIV prevention services? Yes. Or so the data say. In Indonesia, NGOs don't have a monopoly on HIV prevention services for prostitutes. Local departments of health and social affairs run programmes too. It turned out that sex workers in government programmes reported exactly the same levels of condom use as sex workers in NGO programmes: 70 per cent used a condom with their last client. That's far higher than the 50 per cent reported by girls who didn't get any prevention services. No difference in quality, then. Quantity was another story. The government reached 63 per cent more sex workers than NGOs did, even in areas where there was a big NGO effort. In China about ten times as many sex workers said they got condom skills training from government health workers as from NGOs in 2003.[13] In the Philippines most sex workers used government services – and there's virtually no HIV.[14]

Thailand's case is particularly interesting. Prostitution is technically illegal in Thailand, but the industry is vast, characterized by bored women in bikinis watching TV and filing their nails behind plate-glass windows. Sometimes it is one-way glass. A client on the outside can see the girls, but they can't see him. My heart crumples when every now and then a woman stops the nail-filing and makes an effort to send a smile and a seductive look into the ether, not knowing if there is anyone on the other side of the glass to be seduced by it. Clients pick a girl by the number hanging on a plastic disc around her neck, she gets called over the intercom, and off they trot into a back room to do business. It is hard to imagine a less erotic transaction.

In a survey in 1990, 22 per cent of Thai men surveyed nationwide said they bought sex. Young northerners were especially likely to go drinking and tip off to a brothel with their friends. Among twenty-one-year-olds, more than half said they went to prostitutes. No surprise, then, that HIV also ran riot in the north. In the bustling northern city of Chiang Mai, clients carried the virus from one sex worker to another, infecting nearly one in two as early as 1989. Such painfully high infection rates had only ever been seen among drug injectors. Now Thai women were faced with the fact that one in two of the girls who were having sex with their sons, their husbands, their brothers, their policemen were carrying this fatal virus.

The Thai government swung into action.* Thailand's approach was a little more sophisticated than Indonesia's bulldoze-a-brothel-and-build-a-mosque model. And it was more rapidly effective, on a much larger scale, than the 'empower the sex worker to protect her health' approach that has worked well in tightly knit sex work communities such as Sonagachi, in Calcutta. The Thai approach started by recognizing that everyone in the sex industry, from the brothel owner to the prostitute, is in the business for money.

* Unlike most developing countries, Thailand dug into its own pockets very early on to finance huge HIV prevention campaigns. The government installed by the generals after a military coup in 1991 increased spending on HIV tenfold, to US$25 million. Donors topped this up to US$29 million. The government's spending, 86 per cent of the total HIV budget, was an unprecedented proportion for any developing country, an indication of the Thai government's determination to take the epidemic seriously. By 2004 Thailand was accepting a quarter of its HIV budget from donors. Other countries in Asia were sucking in over 85 per cent of their AIDS money from outside sources.

Prostitution may be illegal in Thailand, but it is a national sport, and a very lucrative one at that. Illegal sex and money: the perfect conditions for blackmail. And that is what the government used to create one of the great national success stories in HIV prevention. It is usually referred to by the politer term 'structural intervention' but it is good-hearted blackmail nonetheless. It works like this: Pansak is running a brothel in which people are getting infected with a fatal disease, and he's laughing all the way to the bank. We could shut him down, and turn off his cash flow, just because he's running a brothel. But we won't do that, as long as he shuts down HIV transmission in his brothel. That means making sure every client uses a condom. Pansak's responsibility. Not the girl's, not the client's. Boss-man Pansak's. We're going to screen his girls once a fortnight. If they've got the clap, they're not using condoms. We're also going to ask all the men who come into the clap clinics with drippy dicks which brothels they've been in recently. If they've been in Pansak's brothel, he'd better watch out. We'll treat them, men and women both, but if we have evidence that people are not using condoms in Pansak's brothel, we'll shut him down. No more trips to the bank.

It worked. Condom use in brothels quickly cranked up to 90 per cent. A huge public education campaign got people thinking about the costs and benefits of getting drunk and going whoring. Thai men were not ready to give up sex, but maybe they didn't have to have it in the places where they were most likely to get infected. Some guys started drifting out of the brothels and into the karaoke bars and clubs where they could buy sex from girls who had fewer partners and who were less likely to be infected with HIV. Others just stopped buying sex entirely.[15] HIV fell; so did other sexually transmitted infections.

It was Thailand's government that shut down the HIV epidemic in commercial sex, not its NGOs. Prostitutes were not given much say; they were simply told to refuse to have sex if a client tried to ditch his condom in defiance of the brothel owner, and they had to troop off regularly for their STI screening.

Despite all the calls for 'Leadership', most people in the AIDS industry distrust top-down approaches. The 100 per cent condom-use programme flew in the face of an ideological commitment to letting affected communities lead the charge against HIV. And so we are rewriting history. A WHO report on condom programmes in Burma, for example, says, 'Detailed activities [of 100 per cent condom programmes] may vary from country to country but the main principle is the same: to empower sex workers to refuse condom-free sex services.'[16] 'BOLLOX!!' I noted in the margins of the report. The main principle is to put the enforcement of condom use into the hands of the people with power, the pimps and the brothel owners. Sex workers in organizations like Thailand's Empower may be unhappy about this in retrospect, but at least their members are still alive to argue about it. The programme worked. If we start lying about what made it work, we won't be able to reproduce it elsewhere. Lies are lies, whichever side of the political spectrum they come from. Programmes based on lies don't work.

The Thai programme was an exception. Most HIV prevention services delivered by governments probably wouldn't have existed if NGOs hadn't cleared the path and shown the way. Many of the best government programmes are the ones that have co-opted the NGOs, sucked in their experience, their know-how, sometimes their staff. More and more, governments and NGOs are working together in teams, with one side providing the design tools and the other the production volume.

As we grow clearer about exactly who is at risk, we are in a position to saturate the market of those who actually need HIV prevention services. In most countries, that's well under 10 per cent of adults. We can afford to sacrifice a bit of quality if the trade-off is providing services that are imperfect but good enough for just that 10 per cent.

Peer pressure

The success of the 100 per cent condom programme threw darts at another sacred cow, the 'peer educator'. In plain English, a peer educator is someone who persuades someone else just like them to use condoms and go for health screenings, or not to share needles, or whatever. Peer education worked very well indeed in the gay bars of New York, London, Sydney and Paris. And it has worked quite well in gay bars in other countries, too as Al and his mates in Jakarta showed. The peer education theory was embraced 'almost like the gospel' by the AIDS industry, according to sex work researcher Ivan Wolffers.[17] But it was not so easily exported to other groups. The theory goes like this: a hooker will learn more from a hooker than from some NGO worker or city health official. They speak the same language, they face the same problems, they won't be judgemental. I learned how far this was from the truth when I visited Dongxing in 2004, to review the HIV data and look at the possibility of setting up a 'drop-in' centre for sex workers as part of FHI's programme in the Mekong River region. Dongxing is a dusty Chinese town that squats across the river from Vietnam. The sex scene there is an incredible mish-mash. At the top end are glitzy karaoke bars where pale northern girls in long gowns simper at men in gold jewellery. The cognac flows and the music thuds. In the middle of town are three or four blocks crowded with shops selling one of three things: cell phones, dubious antibiotics and sex (the last two surely connected). The sex shops live a fictional half-life as hair salons – there are barber-shop chairs and the odd blow drier lying around. But you'd be hard pushed to find one with a sink, and I've never seen anyone getting a haircut. All the 'hairdressers' are from next-door Sichuan province, famous for its spicy food and spicy girls.

Along the river-front, looking across to Vietnam, wander a few girls from Hunan, the province which brought us Chairman Mao Zedong and extra-long chopsticks. The chopsticks were allegedly stretched out by poverty – there's so little food on a Hunanese table that everyone has to reach a long way for it. Poorer than the Hunanese and further down the sex-work food chain are the Vietnamese girls who cross the border to sell sex, though usually only on weekends and holidays when business is brisk. And right at the very bottom are the minority Bai-speaking girls who drift into town from the hills, where their tribes predominate. Different languages, different social strata, different motivations. The idea that these women would all want to go to a drop-in centre to play ping-pong together and listen to HIV prevention information seemed a little far-fetched.

'Peer educators' may be sacred to the AIDS industry, but they are not sacred to one another. Prostitutes are rivals. 'I don't want to teach another sex worker how to deal with clients and service them [safely],' a sex worker in Sichuan province told my Chinese colleagues. 'What if she snatches my clients?'[18] The hothouse competition of the brothel does not cultivate people who are secretly longing to work as underpaid social workers. And that's what peer educators are expected to do.

Back in Indonesia, I was wondering about the 'peers are the best educators' orthodoxy while visiting Talenta, a drop-in centre for drug users in Surabaya funded through FHI and staffed mostly by ex-junkies. Well, supposedly ex-junkies. As I sat in the hallway talking to a couple of injectors who had come in to get information on HIV, one of the Talenta staff, Henry, came out of the counselling room. As he saw me, he went rigid. A syringe clattered across the floor and a little bead of blood appeared on his arm, where the needle had dropped out.

I spend my life telling policy-makers that HIV prevention programmes do not encourage risky behaviour, and now our own staff are shooting up in the office, and doing it in front of clients who are being referred to detox programmes. To make matters worse, another Talenta employee was already in jail, accused by the police of using, and we knew the office was being watched. Henry's behaviour threatened the whole programme – I was speechless with rage at his stupidity.

After I had calmed down a bit, I began to wonder who was at fault. We were paying kids who were smackheads this time last year to hang out with junkies every day, and then we were upset when they relapsed. On top of that, we were making the same mistake as we made with sex workers – we were assuming that injectors wanted to help other injectors. One of the favourite catch-phrases in the HIV industry is 'The Community'. This exhausted term assumes that people who happen to do something that puts them at risk of catching a fatal disease feel some kind of solidarity with one another. We don't see that solidarity among sex workers. And we don't always see it among injectors. A trick for getting smack when you're out of cash? Know who's just scored, and point them out to the police. The cops move in and arrest the poor sod, hoping to get paid off for a quick release. As a reward, the 'friend' who turned them in gets some of the drugs they've just confiscated. That's one I learned from an injector in Bandung. Another way to earn cash is to rat on fellow junkies to Granat, a militant anti-drugs group that beats up users.

We've learned from gay men that peer education can work, but it only works in populations where there really is a sense of community. You also need organizational skills to make these programmes work. You can't simply dump some money and a week of outreach training on a group of young people whose life experience is shaped largely by selling sex or taking drugs and expect the programme to succeed. And yet 'peer outreach' is so entrenched as an AIDS industry fashion that that is often exactly what we do.

Think positive

Underpinning all that is sacred in the AIDS business is 'The Community' of people infected with HIV. Their involvement in planning and delivering HIV prevention and care services is a first in the history of infectious disease. Doctors do not wheel out people with leprosy and ask their opinion on how to control the disease. But gay men beat the doctors to it in figuring out what to do about HIV. Doctors who regarded patients as problems began to see that with AIDS, the problems could be part of the solution. These days, patient activism is beginning to scratch away at the monopoly that the medical establishment and its buddies in the pharmaceutical industry have maintained over any number of other diseases. But HIV led the way.

The United Nations, in its inimitable fashion, has managed to co-opt this activism, institutionalize it and reduce it to an acronym: GIPA. GIPA stands for the Greater Involvement of People with AIDS. Everyone, but everyone, now pays lip service to including infected people in the design and often the delivery of HIV prevention and care services.

For care services, this has worked exceptionally well. It's no fun being diagnosed with a fatal disease, especially one that brands you 'Wicked Person'. Being able to talk to someone who has lived through the diagnosis, who has got on with their life, who is still doing an interesting job, who has made peace with their family, who has just made it onto the football team again – that can really help people feel better. Most people with HIV can more or less ignore the fact they are infected with a fatal virus for many years. But when they have to start taking drugs, death casts its shadow over their consciousness at least twice a day. It becomes harder to hide their infection from family or friends once they start popping pills on a regular schedule. It requires a level of organization that many people in their teens and twenties are not used to. I've sat around on many a long evening helping friends count out their pills before a trip. 'Well, I'm going for twelve days, but I'll take an extra four days in case my flights get cancelled. I need four of the blue ones each day, and two of the white. No, no, you do the big white ones, those are only one a day, put them in the square box. I'll put the small ones in the round box. What's four sixteens? 1, 2 . . . 28, 29, 30 . . .' And then the phone rings and you chat to your girlfriend so you lose count and have to start all over again.

People who've been taking antiretroviral drugs for a few years know it is no fun. They can help new initiates with tips on how to deal with nausea and other side effects without just abandoning the pills. They know what it is like to be taking toxic pills every day and have developed strategies that many doctors wouldn't think of. 'Going to the theatre? Put your phone on vibrate and set the alarm for the interval to remind yourself to take your pills . . .'

Helping other people deal with HIV can work well for infected people, too. It can give people a sense of purpose that they didn't have when they were out trawling the nightclubs every evening. Many people have made a career out of being

HIV positive. I remember speaking about this to the Cambodian AIDS activist Pen Mony. Mony dropped out of college because there wasn't enough money to pay for both her and her brother. Her career prospects weren't great, but she got married when she was twenty-one, and she thought all would be well. Immediately, she started to get a lot of sexually transmitted infections. Her husband went for a check-up and was diagnosed with HIV. By the time he died two years later, Mony was infected too. She joined a support group, then started working for the NGO that ran it. She moved up the ladder, learning English on the way. Mony is now in her mid-twenties, extraordinarily beautiful, smart as a whip, open and eloquent. A conference organizer's dream. Over the last couple of years she's flown around the world going to meetings, speaking at conferences, staying in ritzy hotels, raking in the per diems. 'Of course, I would like it better not to be infected,' she told me. 'But it is also true that I have a life now that I could never have dreamed of if I weren't [HIV] positive.' When we spoke, Mony was planning to go back and finish her studies. 'I can afford to now,' she said. Her smile lit up the garden we were sitting in. We chatted for hours, and could have chatted for many more, but her cell phone rang. She excused herself. 'Sorry, that's the driver to take me to the airport.' She was off to the Philippines for three days to represent Cambodians living with HIV at a conference.

Being infected with HIV does not glue people together if their backgrounds are too disparate. Mony said that support groups for HIV positive women in Cambodia often broke down because infected housewives didn't want to talk to infected prostitutes. 'If a woman starts telling her story and it is clear she was a sex worker, you can see the faces of the others change. They blame [the sex workers] for infecting their husbands. So they think: I'm infected because of her,' she told me.

So the assumption of 'Community' doesn't necessarily hold for people who are infected with HIV, any more than it does for sex workers or drug injectors. People who have great communication skills and who happen to be infected with HIV themselves will make good counsellors for people who test positive. But it is the communication skills that make them good counsellors, not the HIV. The reverence paid to people with HIV in the AIDS industry stems in part from the fact that so many of them have put valuable skills to good use in damping down the epidemic and cushioning the blows it deals to the people it affects. Without those skills, the reverence would be misplaced.

In the AIDS industry one does not speak critically of an HIV positive person involved in 'The Response' to the epidemic, especially if one is part of the biomedical mafia. It smacks of a desire to undermine the achievements of the activists, without which the response to HIV would probably be languishing with the response to leprosy and dengue fever. But I feel the need to whisper on behalf of Joyce. Joyce was married to Charles, who worked for me when I was commuting between Geneva and Nairobi. When she got pregnant with their third child, she and Charles decided to take HIV tests. There weren't many options in Nairobi in 1999, so I took them up to the counselling and testing centre run by TAPWAK, the Association of People With AIDS in Kenya. When I picked them up after they got their results, Joyce was shaking with terror. Charles did his best to comfort her, but no chance. Joyce had tested HIV positive. Charles had not. I added my soothing words to Charles's, promising that we'd get drugs to reduce the likelihood that her baby would be infected, promising to find her a decent doctor. Once Joyce was safely home, Charles explained her distress. The diagnosis, obviously. But also the way it was delivered. The counsellor, a woman who was visibly in the advanced stages of her HIV disease, had gone on a Sodom and Gomorrah rant about the infection being fair punishment for Joyce's wickedness. 'She pointed at my wife and she was shouting,' said Charles. Usually completely unflappable, even he had been upset by the experience. 'She said: "It's your own fault." She said: "You think you are so pretty, but you'll see, soon you will be a death's head, just like me."'

Being HIV positive does not necessarily make you a good counsellor, any more than being fat makes you a good dietician. HIV is a virus, not a job qualification.

6

Articles of Faith

The job qualifications of people with HIV may occasionally be questionable, but their commitment rarely is. People who work in AIDS prevention and care because they and their closest friends have been affected are driven by a missionary zeal. They are generally passionate about improving the lives of people who have HIV or who are threatened by it; and many of us draw great strength from that passion.

We need that strength to confront missionary zeal of another sort. Religious fervour, especially Christian and Muslim fervour, stands firmly between what we should be doing to prevent HIV and what we actually do. It obscures the truth, and directs the HIV prevention industry down one dead end after another. We trot along as directed because power follows religion, money follows power, and we follow money.

I'll go back to the money later. But I'd like first to look at how religious dogma crushes our efforts to translate good data into good HIV prevention.

Here's wisdom from President George W. Bush: 'We need to tell our children that abstinence is the only certain way to avoid

HIV. It works every time.'[1] He did not add, 'when used consistently and correctly', the health warning that always pops up when we talk about how effective condoms are in preventing HIV. Yes, abstinence works every time, if kids really do abstain from both vaginal and anal sex all the time. But of course they don't.

Several hundred million years of evolution have loaded adolescent bodies up with hormones that drive people to have sex. Sperm costs a body far less to produce than eggs, so men have a greater incentive to spread their reproductive capital than women do. That means men are more likely to try and clock up lots of sex partners. But they also want to try and keep other men away from the women most likely to reproduce their own genes, i.e. their wives. As human societies evolved, they tended to press women firmly into the mould of chastity and fidelity, while accepting that boys will be boys. But if women are all battened down, who will the philandering men sow their oats with? Enter the sex worker. While society likes to think of prostitutes as 'wanton women' (or men), some sex workers see themselves as guardians of virtue.

'What I do is social work, really,' said Fanny, a waria who works under the bridge near the Italian Cultural Centre in Jakarta. Fanny's social conscience is so strong that she often waives payment for schoolboys and college kids. 'At that age they are randy as hell, and they're always broke. So you give them a freebie, just to be nice.' 'If they're cute . . .' observed another waria who was squatting under a nearby streetlamp repositioning a false eyelash that had gone walkabout across her face. 'In their teens, they're all cute,' laughed Fanny. An opinion handed down from the ripe old age of twenty-four.

Sex with a prostitute is an age-old solution to the problem of young male lust in many cultures, one that doesn't carry the risk of impregnating some 'nice' girl or shaming a good family. Young female lust has been dealt with in many societies by early marriage – the risk of pregnancy a very strong deterrent to any premarital experimentation. Now contraception is eroding that risk. More importantly, girls have now got better things to do than get married at fifteen. All over the world, more and more girls are staying in school, going to university, getting jobs. Their social horizons are changing. Their hormones are not.

Around a fifth of young Americans tell researchers that they have been through a ritual 'Virginity Pledge', signing a contract that declares they will not have sex until they get married.[2] They do it at huge religious rallies and concerts, they do it at high school ceremonies, they do it at glitzy balls in spa hotels. The weirdest virginity pledges are signed not by kids but by their fathers, at those glitzy balls, where girls in floaty white dresses with see-through skirts slow-dance with Daddy. Before the assembled crowd, Daddy pledges 'I, (xyz)'s father, choose before God to cover my daughter as her authority and protection in the area of purity.'[3]

Pledgers put on silver rings to signal their commitment. They fill in a little 'virginity' wallet card, that they can carry to remind themselves not to stray. And then they have sex.*

A study that has followed over 11,000 US adolescents across several years found that 72 per cent of pledgers had sex before marriage. Pledgers did wait a bit longer before they first had sex, and did have fewer partners than non-pledgers. But they

* They have sex despite supportive websites which suggest '101 things to do (besides having sex)' – though I suspect they mean (instead of having sex). At number 2 is 'Go shopping at the mall with your friends.' Further down the list are: talk on the phone, make a scrapbook, visit a nursing home. Have an orgasm or visit a nursing home? Now let me see . . . (See www.iamworththe-wait.org/101.html), accessed 19 June 2007.) If you'd like a handy pledge card for your wallet, you can also download one from this site.

were also less likely to use condoms when their hormones got the better of their intentions, and they were just as likely to have sexually transmitted infections as kids who never swore off sex.[4]

Even on their home turf, abstinence-only programmes fail.[5] Do US authorities look at the evidence and think: hmmmmm, maybe we should try something else? Far from it. They are busily exporting the programmes around the world.

The United States is by far the most generous funder of AIDS programmes for the developing world, in raw cash if not on a per-taxpayer basis. As we saw, the Christian lobby, which has been inordinately influential in the court of Bush the Younger, regarded AIDS as an abomination until their eyes were opened to the havoc it was wreaking among the innocent wives and babies of Africa. Once they 'got religion' on AIDS, they were swift and effective in pulling strings to increase spending on AIDS in Africa.

At the start of 2003, Bush and his advisers were bristling for a fight in Iraq. The war-mongering had already provoked protests; bloody inconvenient in cities like Jakarta, where thousands of students rushed to the streets, clogging up arteries already choked with traffic. Their rage was palpable. Across the country, and in countless other countries across the globe, there was a feeling that Bush was prodding Americans towards xenophobia. The White House needed to prove that this was not so. Those lobbying for more spending on AIDS in Africa provided the right compassion at the right time. And spending on AIDS would prove popular with the pharmaceutical companies that would supply the medicines, as well as with the Christian groups. All useful in a pre-election year.

Bush embraced what he called 'a work of mercy – to help the people of Africa' and made it his own. The result was the US$15 billion 'President's Emergency Plan for AIDS Relief' (PEPFAR).[6] Not the US taxpayers' plan. The President's Plan.

Bush's Christian courtiers worked with the staff of conservative Congressmen to massage their interests into the law that authorized the PEPFAR spending. They did well. Over half of the money was to be channelled through NGOs, and faith-based organizations were singled out for special attention.[7] The glorified position given to People Who Pray was sometimes comical. A list of indicators we were supposed to use to report all the wonderful things PEPFAR was doing included: 'Number of faith-based laboratories that can conduct quality HIV and STI testing.' I don't think anyone really thought that lab technicians who pray are any better at reading test results than anyone else. But all the early indicators had their faith-based component, because PEPFAR's administrators wanted to sing the song that Congressmen wanted to hear.

'We will make policy decisions that are evidence-based,' the President's Emergency Plan declares. But by law, 20 per cent of the PEPFAR money must be spent on HIV prevention, and one-third of that is specifically allocated to programmes that do nothing but push abstinence until marriage.[8] That is US$1.06 billion to fund foreign programmes that have a failure rate of 76 per cent even among the American kids who choose to sign on for abstinence. Among those who sign on for consistent condom use, the failure rate is 2 per cent.*

* The bill was passed by a Republican-controlled Congress. Many people hope that this provision, along with other restrictions, will be dropped if a Democrat-controlled Congress authorizes another $30 billion in 2008, but if domestic policy is any guide, the smoke signals are not good. In June 2007 a committee of the Democrat-controlled Congress voted to increase federal funding for community-based abstinence-only programmes in the US to nearly $170 million in 2008, from just over $140 million in 2007. For citations on abstinence see Chapter 5; for condom effectiveness see this chapter note 20.

New maths

As the age at marriage rises, the likelihood of virginity on the wedding night drops. I wonder how many of the American kids pledging virginity until marriage do the maths. To keep the pledge, the average American guy would have to go through all of his hormone-charged teens and four-fifths of his twenties without ever having sex. Girls could be a little less patient, but they'd still have to forgo sex until twenty-five.[9] Frankly, that's not what most people dream of while they are young, energetic and as sexy as they will ever be.

Even people who are not optimistic about the success of abstinence-until-marriage programmes tend to believe that an increase in premarital sex will logically lead to an increase in HIV. In many of the countries where I've worked a lot of girls have traditionally kept their legs crossed until they get married. Asian women top the list for virginity at marriage, followed by Latin Americans (though young Brazilians get around more than young people elsewhere on the continent). There's not much HIV in the general population in either of those continents. In Africa, there's lots of premarital sex, and lots of HIV. Everything is following the more premarital sex = more HIV pattern so far. But look at industrialized countries. In Europe, Australia, Canada and even virginity-pledge-prone USA, almost everyone has sex before marriage but very few heterosexuals have HIV, unless they shoot up drugs or sleep with someone who does.[10]

So is it necessarily a bad thing if young people in Asia or Latin America become more like young people in the West? The shift does seem to be taking place; I can even see it at my dinner table in Jakarta. When I was first in Indonesia in the late 1980s lots of my friends were quite a bit older than me, and many of them had school-age kids. When I came back a decade later, those kids had blossomed into young adults; it was this group that used to come and sit around my dinner table, clogging up the ashtrays, slapping absently at mosquitoes and gossiping late into the night. Topics of conversation were the universal favourites – movies, clubs, cars, cell phones, sex. Why aren't I getting more of it? Have you tried those new mango condoms? What?! Your mother doesn't let you sleep with your boyfriend when you're at her house? Oh please . . .

Everyone had tales to tell. Nungky, the not-yet-thirty architect who turned up in her mum's Merc to keep it away from the junkie brother, used to go out with Luwi who was great mates with Alex who had been with Lisa for five years. Lisa was best friends with Renjani, who (besides being the sister of Luwi) lived with Paul, more or less next door to Lisa and Alex. Then Renjani dumped Paul and went off and got pregnant by Ashley, while Nungky dumped her boyfriend Luwi and started going out with Mitu, who (besides being a woman) was an old family friend of Luwi and Renjani. This left Luwi despondent, until he started sleeping with Lisa, which put Alex into a fury, and suddenly everyone was either sleeping with or hating everyone else, and inviting people to dinner got very complicated indeed.

My friends were not typical of all young Indonesians. But these sorts of soap operas are no longer particularly unusual among middle-class urban kids in Asia, of whom there are tens of millions. In cities all over East and South-east Asia, young women are claiming the right to a sex life, and once they get started, they seem to be making up for all the time lost by their mother and grandmothers. In a Japanese survey in 1999, around 40 per cent of girls aged 18–24 had racked up five or more partners, and about the same percentage said they had had sex with two or more men just in the past year. Don't think this is only happening in the neon-lit streets of the cities. In rural high schools, 30 per cent of both girls and boys said they had had sex, and over half of them had had more than one partner.[11] An online survey of women's sex lives in China drew 32,000 responses, 15,000 of them from single women. Predominantly young, urban and college-educated, nine out of ten said they enjoyed sex, and nearly one in five pronounced herself 'extremely satisfied' with her sex life.[12] Online surveys are problematic for many reasons; even if the respondents are who they say they are the figures will be skewed because women who haven't had sex or don't much like it are less likely to answer the survey. But still, there's no smoke without fire. The survey does rather cloud up our clear view of sex in Asia, a view which assumes that 'nice girls' don't put out until marriage. This line was promulgated by senior Chinese officials until well into the late 1990s.

I took issue with one of these officials when he said extramarital sex was 'not in our culture'. It was at a meeting in Geneva in 1997, when the higher-ups at UNAIDS were trying to persuade the Chinese government that it should take HIV seriously. An esoteric undergraduate degree in classical Chinese came into its own for the first time since I'd graduated in 1984. I gleefully plunged back into one of the texts I had studied at Oxford, the *Golden Lotus* – a sprawling sexual soap opera dating from the Ming Dynasty (1368–1644). It is about a nouveau riche merchant who bribes local officials to get lucrative contracts, keeps four wives, numerous concubines and a stable of female and male prostitutes. With so much competition, none of the wives are getting enough sex, so they top up by having affairs with the gardener, the rent boy and the beautiful young cousin of the tailor's wife, or some such.

I drew a little 'sex map' of the characters in the novel, and even colour-coded what we would now consider the high- and lower-risk sexual partnerships (gay sex and prostitutes high, occasional non-commercial lovers lower, spouses with no commercial or gay partners lowest). I used the generic map in a training exercise to demonstrate how widely HIV might spread if you dropped the virus into prostitutes or the gay community. And again from the Chinese official I got 'Yes, but in China we don't have to worry about such things. It is not in our culture.' Then, meanly, I told him where the map came from. He looked blank for a moment. The *Golden Lotus* is banned in China, to preserve morality, though most people know about this gaping hole in the literary canon. When I jogged his memory, he blushed. Later, one of his junior staff came over to ask if I had a copy of the book they could borrow.

In Jakarta, Tokyo, Beijing or Bangkok, it seems young Asian women are having more premarital sex than ever before, and quite a bit of it is condom-free. This has allowed some people to gloat ('You see! We knew MTV and Hollywood movies would corrupt our youth') and made some people panic ('Oh no! If the kids are all having sex, that means HIV will spread like wildfire').

There's no evidence that schmaltzy pop videos and crap films contribute to people having more sex. But what is the evidence that more premarital sex translates into more HIV?

Let's do the maths using the Northern Thai data. Thailand used to fit the classic 'virtuous girls, philandering boys' model. At the start of the 1990s, 57 per cent of twenty-one-year-old men in Northern Thailand trooped off to the brothel to do their philandering. More than half the sex workers who soaked up their excess energy were HIV-infected. Just over 60 per cent of the young men used condoms with these prostitutes. So overall, a minimum of 13 per cent of twenty-one-year-old guys in Chiang Mai would have had unprotected sex with an infected prostitute in 1990. Less than a quarter of young men had girlfriends, and only one guy in every 500 would be exposed to HIV in non-paying sex. Altogether the risk of exposure to HIV would have been 13.2 per cent.

Then two things happened. Thailand expanded its famous 100 per cent condom programme, and the Thai economy boomed. Girls were getting better educations than ever before; in fact a daughter was slightly less likely to drop out of secondary school than a son.[13] Educated girls were waiting longer before getting married, but not before having sex. By the end of the 1990s, 45 per cent of girls aged 15–21 in northern Thailand admitted to having sex with boyfriends before marriage, compared with less than a tenth of that in a nationwide survey in 1993. Boys reported more girlfriends too, and everyone reported lowish condom use – lower, certainly, with a girlfriend than with a prostitute.[14]

So at the end of the decade, we have a lot more premarital sex and not all that much condom use with girlfriends. But now that these young, cash-strapped guys can have sex without paying, they've stopped handing over cash for sex. By the end of the 1990s, only 7 per cent of young men were paying for sex, and HIV prevalence in sex workers had come down too. If we do the same sums, we find that altogether, 1.4 per cent of uninfected young men in Chiang Mai were at risk of being exposed to HIV.[15]

So the proportion of guys reporting sex with girlfriends nearly doubled, while the risk of exposure to HIV fell by at least 80 per cent. The risk of a new HIV infection among young men in Chiang Mai would have fallen by far more than 80 per cent, because girlfriends are a lot less likely than sex workers to be infected with one of the other sexually transmitted infections that open the door to a new HIV infection.

In short, more women having premarital sex equals less HIV.

In countries where prostitutes have always protected the virtue of 'nice girls' and mopped up the needs of young men, a shift to sluttish 'Western' sexual habits is not something to panic about. I'm not expecting governments to rush out with programmes to promote premarital sex. I think economic development will do that for them. As Asian and Latin American countries grow richer, as girls spend longer in school, and as there are better jobs for them to do when they graduate, women will get married later, and have more boyfriends before they tie the knot. And that means guys can have more (unpaid) girlfriends, so both boys and girls will be happier, and norms of commercial sex and the risk of HIV transmission will fall.

Sloshing more than a billion dollars into exporting failed abstinence-until-marriage programmes for young heterosexuals is not enough for Bush's Christian courtiers. Abstinence is for everyone who is not married. Before Lenny first sent her team of waria on to the streets of Jakarta to cajole their peers into using condoms with clients, 43 per cent of waria said they used a condom the last time they sold anal sex. Lenny and her team did such a good job that condom use rose to about 80 per cent over the next two years. Most salespeople would be congratulated for those results, but under US regulations we were supposed to tell Lenny and her team that their sales pitch was wrong. They were supposed to start it with: 'The only sure way to avoid HIV is to avoid sex.' If you sell sex to earn your living, avoiding sex is also your best bet for avoiding paying your rent, for going hungry, for not being able to buy medicine. But if an HIV prevention programme for prostitutes is paid for by the US taxpayer, we are supposed to recommend unemployment as the first line of defence. The rules are that abstinence must be part of every programme to prevent the sexual transmission of HIV among people who are not married, everywhere in the world.

Organizations like Lenny's often simply ignore those of the funders' rules that get in the way of doing good programmes. The professionals at USAID, CDC and the other organizations that are handing out the cash realize this. But most of them care more about preventing HIV than about rules that pander to far-away political interests. If they can't get the rules changed, they'll close their eyes and let people like Lenny get on with what works as much as possible. They get in trouble, though, if the courtiers in Washington get wind of a bending of the rules.* I'm frankly amazed that so many people within organizations like USAID manage to go calmly on trying to do the right thing when the odds are stacked so heavily against them.

Under cover

The increase in premarital sex and decrease in prostitution is a pretty good example of how development – especially more opportunities for women – does, in fact, cut the risk of HIV. But I still maintain that we can't be sitting around waiting for development to take care of the problem. Thailand has taught us that fewer men buying sex translates into lower risks for HIV infection. But it has also taught us that it is easier to get condom use up than keep the sex industry down. The proportion of men buying sex halved over four years in Thailand. The proportion not using condoms when buying sex halved in just a year and a half. As condom factories get busier, the workload in the VD clinics falls. Condom use in commercial sex rose sevenfold in

* See, for example, the acidic letter to James Kunder, USAID's Asia and Near East Bureau boss, from conservative Congressman Mark Souder. Souder felt Kunder was being soft on an Indian NGO that gave back money to USAID rather than comply with rules set by Washington: www.genderhealth.org/pubs/SANGRAMSouderletter100605.pdf.

Thailand in the early 1990s. Patient numbers at VD clinics fell sevenfold in the same period.[16]

Where condoms have been promoted without let or hindrance, they tend to get used, especially in the riskiest encounters. In Thailand and Cambodia, in Vietnam and China, in Nepal and the south Indian state of Tamil Nadu, in Brazil and Australia, in the Netherlands and Britain, condom use in commercial sex is the norm, and condoms are very frequently used in casual partnerships too. And where condoms are widely used, whether it is in the brothels of Thailand and Cambodia or the bedrooms of San Francisco, they can prevent and even reverse the HIV epidemic.

Despite overwhelming evidence that this is the case, opposition to condom promotion persists.

I'm in another four-star hotel with marbled lobby and musty carpets for another expensive pow-wow about HIV. It's 2003, and after much prodding, the Indonesian government is finally launching a National Anti-AIDS Movement. Because AIDS was being sold as everyone's problem, everyone had to be there. There were the women's association and the youth forum, the transport workers' union and the education department. There was even the odd researcher. I was assigned to a round table on condom use, chaired by army general Kiki Kilapong, the one who had the dud HIV test kits dumped on him. My colleague Chris Purdy, who started a fashion by introducing condoms that taste like the stinky durian fruit, had just given a talk about condom sales: rising, but still short of the market potential. At the time, we estimated that around 8 million Indonesian men bought sex fairly regularly, accounting for around 65 million sex deals a year. Fewer than one client in ten said he used a condom every time he went to a hooker.

'I'm very happy to hear that!' pronounced a Muslim cleric,

who had been fidgeting and tugging at his beard in agitation during Chris's presentation. There was a stunned silence. 'It's obvious that condoms are evil,' the cleric ploughed on. 'So if people don't use them, it proves that we are a religious nation.' This guy, with his tidy skull-cap and his flowing robes, apparently has no problem with you going to a hooker, picking up a disease, taking it home to your wife. As long as you don't use a condom, you're a good citizen of the religious nation. It's quite usual in Indonesia to focus on the letter of the religious law rather than on its spirit. I've met prostitutes who have been married hundreds of times. Good Muslim clients perform the wedding ceremony for themselves before starting in on the girl. An hour later, they divorce her. Since they were married while they were having sex, they have not sinned. But this cleric's 'condoms are evil' thing doesn't even conform with the letter of the law. If anything, the Islamic scriptures tend to support the limitation of family size. The demonization of condoms, now fairly common in Muslim countries, seems to be a relatively recent contaminant from Catholic doctrine.

The Vatican clearly believes condoms are the work of the devil. And they are perfectly willing to claim that black is white to dissuade people from using them. 'The AIDS virus is roughly 450 times smaller than the spermatozoon,' Cardinal Alfonso Lopez Trujillo, president of the Vatican's Pontifical Council for the Family, told the BBC's *Panorama* programme in 2003. 'The spermatozoon can easily pass through the "net" that is formed by the condom.'[17] This man, who presides over the Family Council of a state with no women and no children, believes that sperm can pass through a condom. He may also believe that a camel can pass through the eye of a needle, but there has never been a recorded instance of either thing happening. Condoms occasionally burst – up to 10 per cent of the time if they have been stored badly or are used incorrectly – and then anything can get through them.[18] But properly used, a condom holds water, it holds semen, it holds sperm and it hold viruses. Fewer than 2 per cent of heterosexuals who are infected with HIV pass the virus on to their long-term partners if they always use condoms, compared with 10 to 18 per cent of those who don't always use condoms.[19]

Another piece of anti-condom doctrine holds that telling young people about sex, and especially telling them that they can have sex without getting sick or pregnant if they use condoms, will make everyone rush to drop their knickers. Here's Ndingi Mwana a'Nzeki, the Archbishop of the Kenyan capital Nairobi, on the subject, in 2006: 'There are no two ways about it . . . When condoms are provided anyhowly, chances of promiscuity increase since a majority of our people end up engaging in casual sex.' In the most recent national survey in Kenya, one in five unmarried women over the age of fourteen reported having sex, and twice as many unmarried men said they were getting it. Two-thirds of that sex was unprotected, so it seems unlikely that it was the availability of condoms that made them have sex. The Catholic big shot was speaking during a conference on stigma, discrimination and denial.[20] And perhaps he knew something about denial. As he was urging the government to ban the advertising and distribution of condoms, HIV was already destroying the immune systems of one in seven people in their late teens and early twenties in some provinces of Kenya.

The cardinal was implying that when condoms are available, you can get laid any time you like. Sadly, there is not a scrap of evidence to support this view; it is a bit like suggesting that you can end a drought by giving out umbrellas. There is, however, plenty of evidence to suggest that easy availability of condoms increases the chance they will get used between people who were going to have sex anyway. In sex establishments that have condoms available in Indonesia, condom use is 50 per cent higher than in places where clients have to go outside to buy a condom. Girls who have condoms to hand when they are interviewed are more than twice as likely to use condoms than those who don't.

The thing about faith, about doctrine, about ideologies of any sort, is that you can't fight them with facts. If someone *believes* that wearing an amulet will protect them from the Evil Eye, or that they can make a neighbour sick by sticking pins into a wax doll, if someone *believes* that condoms are inherently evil, there's not a damned thing I, as a scientist, can do about it.

The Vatican has put ideology before science for centuries. More recent practitioners of the dark arts of data manipulation can be found in the administration of US President George W. Bush, according to the government's own Surgeon General, Richard Carmona. Carmona was the first person Bush appointed to the post of Doctor-in-Chief, in 2002. He quit in disgust four years later, having found that he was supposed to do nothing more that rubber-stamp the opinions of political appointees with no scientific training. 'Anything that doesn't fit into the political appointees' ideological, theological, or political agenda is ignored, marginalized, or simply buried,' he told a congressional committee in July 2007.[21]

'Ideological, theological, or political agenda,' Carmona said. More and more, it seems to me, these three things are merging into a single sword of Righteousness.

Carmona's statement surely came as no surprise to Henry Waxman, an extraordinarily energetic Democratic Congressman from California who chairs the Committee on Oversight and Government Reform. (Oversight? Misspending billions of dollars of taxpayers' money on programmes that we know don't work is an oversight?) 'Simply put, information that used to be based on science is being systematically removed from the public when it conflicts with the Administration's political agenda,' Waxman said in a letter to the US Secretary of Health and Human Services, Tommy Thompson.[22]

Waxman and thirteen other Congressmen were railing against a particularly perfidious piece of right-creep which refashioned the Centers for Disease Control into a turncoat against condoms. CDC, America's public health agency, has done fantastic work supporting HIV prevention in the United States, and has a growing presence overseas. Since September 1999 the CDC website had been home to a very informative fact sheet about condoms. It included hard data from studies of condom effectiveness so that people could judge for themselves how protective condoms might be. Answer: very. It included studies that investigated whether condom promotion made young people rush out and have sex. Answer: no. Quite the opposite, in fact. It included explicit instructions for condom use. It included this statement: 'For those who have sexual intercourse, latex condoms are highly effective when used consistently and correctly.'[23]

In January 2003 the fact sheet disappeared from the site. Its replacement also masqueraded as a fact sheet, but most of the facts were gone. The data on condom effectiveness, the studies showing that sex education delays sexual activity, the instructions on how to use a condom – all gone. And now, in bold text, it included this statement: 'For persons whose sexual behaviors place them at risk for STDs, correct and consistent use of the male latex condom can reduce the risk of STD transmission. However, no protective method is 100 percent effective, and condom use cannot guarantee absolute protection against any STD.'[24]

The new wording is still technically correct. But it doesn't exactly fill one with confidence. Old message: condoms work.

New message: condoms might work some of the time for wicked people who choose to take sexual risks.[25]

Then there are inept misinterpretations of the data. Bill Bennett is a talk show host who shares his conservative opinions with millions of Americans every day in a radio show called *Morning in America*. Commenting on a report that African countries that have the highest rates of condom use also have the highest rates of HIV, Bennett came up with this gem: 'Clearly, condoms must no longer be considered the first line of defense against HIV prevention.' That's like saying hospitals have more sick people than schools, so we should no longer consider hospitals an important part of the health system. Bennett might be expected to treat data with greater respect – he was America's Secretary of Education in the days of President Bush the Elder.

For my money, people have a right to choose any ideology they want, and to live by it to their heart's content. Think condoms are evil? Fine, don't use them with your one, lifelong, mutually monogamous partner. But don't undermine the science that persuades other people, people who don't have one, lifelong mutually monogamous partner, to use them.

American conservatives clearly look to dogma before science when thinking about sex and condoms. And they are exporting that dogma, sometimes in subtle ways. PEPFAR, for example, allows for condom use 'as appropriate'. It clarifies: couples may use condoms if one spouse is infected and the other isn't. It reinforces this restriction by saying repeatedly that people cannot take steps to prevent HIV unless they know they are infected. Of course that's nonsense. You don't need to know you're infected with HIV to use a condom, a clean needle, or lubricant. You don't need to know you're infected to get screened for other STIs and have them treated. In most of the world we know perfectly well who needs prevention services. We don't need to test anyone to be able to provide wholesale HIV prevention services for the small fraction of the population that actually needs them. But now UNAIDS has taken up the 'prevention's impossible without testing' refrain.*

It's easy to get cross with the anti-condom ideologues who wielded their pens over PEPFAR. But don't imagine that their messages are unwelcome, that, if it were not for US prudery, other countries would be promoting condoms more energetically. Condom use has stayed stubbornly low in commercial sex in Bangladesh, Pakistan and Indonesia.[26] And, for a shamefully long and epidemiologically devastating period, in the countries of East and Southern Africa. What do those countries have in common? In the Asian countries, Muslim groups wield considerable political power. In many of the African countries, Christian churches have a loud voice in public affairs. Here are some developing countries where condom use has sky-rocketed: China, Vietnam, Cambodia, Thailand, Brazil.[27] Two Communist, two Buddhist and one the home of liberation theology.

In Indonesia, the cleric at the conference was not alone in his distaste for latex. The previous year FHI had worked with the health ministry to develop a campaign to increase condom use among men who visit brothels. It was aimed firmly at the lower-middle class men who our research suggests are most likely to

* PEPFAR says: 'Without knowing their [HIV] status, individuals can . . . not take steps to prevent transmission to others.' And this from the UNAIDS website: 'The vast majority of people living with HIV in the world are unaware of their status, leaving them unable to . . . change behaviours that might put themselves and others, such as sexual partners, at risk of (newer) infections.' www.unaids.org/en/Policies/Affected_communities/PLWHIV.asp, accessed 20 June 2007.

be trawling around picking up sexual infections. The ads featured ageing Indonesian pop icon Harry Rusli chatting at a drinks stall. In another risqué scene a doctor in a white coat talks about condoms as he puts the cap firmly on his pen. Very suggestive. The ads ran for a couple of weeks in 2002. Then the Indonesian Mujahiddin Council (MMI) wrote to the TV stations that were airing it, saying that they could not be held responsible for what might happen if the ads continued to run. Well, of course not. Fundamentalist religious groups aren't responsible for TV stations. But MMI was led by Abu Bakar Baasyir, who *was* held responsible for the bombings which killed 202 people in a nightclub in Bali later that year. Baasyir was well known as a firebrand, and the TV stations crumbled in the face of his illegal threats and pulled the ads. The health ministry, whose logo was on the ads, did nothing to try to enforce their service contracts. In the world's largest predominantly Islamic democracy, militant Muslims had wiped out a campaign designed to save lives. A victory for public health terrorism.

A bit of railing against a condom campaign can be a good thing. Chris, the inventor of stinky durian condoms, reminded me that when Cardinal Sin of the Philippines vented about a new condom campaign, his comments were splashed across the front pages of the newspapers. Condom sales in Manila were catapulted upwards. Chris has recently been plastering Indonesia's red light districts with posters promoting Sutra brand condoms. The Kama Sutra campaign includes some mildly exotic sexual poses.[28] 'We were half-hoping for a protest or two but wouldn't you know it, not a peep!' he wrote to me. You want a bit of fuss, but not so much that the campaign gets shut down.

Religious leaders say all sorts of things about condoms. Scientists say they work. A review of studies among couples where one person is infected and the other is not found that the HIV-negative person was five times more likely to get infected if they did not always use condoms with their partner than if they said they did.[29] The data suggest condoms are not 100 per cent perfect – they do burst, infrequently, and people do use them carelessly. But they cut your risk of infection by at least 80 per cent, without your having to cut down on sex. What's not to like?

The real truth about condoms

There's quite a lot not to like, actually. The truth, the whole truth and nothing but the truth: sex without a condom is more fun than sex with a condom. So help me God, but it really is true.

Condom thought bubbles:

'Wayhay! Looks like I'm in there! Have I done enough of the foreplay thing yet? Where's my condom. Oh damn, it's in my jeans pocket' – *vision of jeans peeled off and discarded behind the sofa.*

'This is nice, but looks like he's getting impatient. Where's my condom? Oh damn it's in my jeans pocket' – *vision of jeans shucked off on the bathroom floor.* 'Hang on, maybe there's one in the bedside table. Hope he doesn't think I'm a slut.'

'Great, she's got one. The slut.'

'Great, I've got one. Now to get it on him. Oh damn, I wish they'd colour-code these things. Which is the inside? Oops.'

'God, I wish she'd get on with it.'

'I'll try it the other way. Oops. No, I think I was right the first time. Well, at least he won't think I'm a pro.'

'Well, it doesn't look like she does this too often. Wonder what the expiry date on the condom is . . .'

Condoms interrupt the flow of sex, they create embarrassment, they reduce sensation for both men and women (though nothing like as much as the 'like-taking-a-shower-with-your-raincoat-on' brigade would have us believe). They also protect you from pregnancy, which may or may not be a good thing. They protect you from a range of unpleasant but curable diseases (gonorrhoea, chlamydia, trichomonas and syphilis, to name the most common), as well as from unpleasant and incurable diseases (herpes, hepatitis). Not to mention an unpleasant and early death (HIV). Embarrassment versus death. You choose.

Of course I can't choose, because I am a woman. This is held as an article of faith not by any church, mosque or temple but by the AIDS establishment. It is propagated with religious fervour by those who have swallowed whole their own rhetoric about HIV being spread by poverty and gender inequality. Virtually every single discussion of condom promotion contains some statement about women's inability to protect themselves. 'Economic and social dependence on men often limits women's power to refuse sex or to negotiate the use of condoms,' says the UNAIDS fact sheet on women. The implications of this are two-fold. One: if women could use condoms they would because, two: women care if they die, whereas men don't. I draw this second conclusion because men obviously can use condoms – they have the penis and the power. If men don't use them, it must be because they don't value their lives.

I think both implications are hogwash. Men don't use condoms for lots of reasons. Reasons one, two and three are that sex is more fun without condoms. You can see the cost-benefit analysis as embarrassment versus death. Or you can see it as more fun now (a sure thing) versus more pain later (uncertain, since you may be able to have unprotected sex and get away without being infected anyway). You choose. Men often choose the fun. So do women. Look at the female condom; women can choose to use that in lots of countries. Most don't. That may be in part because it looks like a supermarket shopping bag stuck up your pussy, with handles hanging out the bottom. Or because it sounds as if you're making love inside a packet of potato crisps – crinkle-crackle-crinkle-crackle. Which of course makes you laugh, but you must never laugh when you've got a female condom in because you can feel that rubber ring scrunching up inside you and ouch that hurts and suddenly it isn't so funny any more and if you can't laugh, what's the point of having sex?*

The cost–benefit analysis is very different for different people. A woman whose husband trawls the brothels has a greater incentive to use condoms than a college kid who has picked up the star player on the football team. A man who knows his wife used to shoot up drugs has a greater incentive to use a condom than someone who marries the village school teacher. Does this sound stigmatizing or discriminatory? Perhaps, but facts are facts. Some people are more likely to have been exposed to HIV than others. People can and do weigh up these facts in their own lives.

Sex workers have a pretty high incentive to persuade clients to use condoms. The conventional wisdom embraced by many in the HIV prevention industry is that they can't choose because their bargaining power is so low. But chat to a sex worker or look at the data and you'll find that's not always true.

Let's go back to Nana, the Chinese-Indonesian girl who sells sex in the booming back-alleys of Rawa Malang. She doesn't

* For a more balanced view see Adu-Oppong et al., 2007; Okunlola et al., 2006; Telles Dias et al., 2006; Thomsen et al., 2006; Welbourn, 2006; Witte et al., 2006. Or just try one. But don't laugh.

find it so hard to get clients to use condoms. 'The thing about men is that they're lazy,' said Nana. 'They're lazy about buying condoms or suggesting we use them, but once a guy is standing there in front of you all hot and bothered and ready to play and you put a condom on him, he's certainly not going to fight. Too lazy.'

The stats seem to confirm Nana's experience. Just over a third of some 10,000 female prostitutes across Indonesia said they had asked all of their clients in the past week to use condoms (higher than I would have expected, though lower than I would have liked). Does asking translate into using? Very often, it seems. Women who ask all their clients to use condoms used them fifteen times more often than women who didn't bother asking.[30]

If all a sex worker has to do to get a client to use a condom is ask him to, why do so many not bother? The conventional wisdom: because they get paid more for 'skin to skin'. This is supported by 'mystery client' surveys, where researchers pose as clients and negotiate, hard, for sex without condoms. A lot of the time, sex workers eventually give in. Although not always. An acquaintance of mine suggested unprotected sex to a Mongolian prostitute working out of a well-known Beijing club. In the interests of research, of course. She slapped him for his audacity, and he was asked to leave the club.

But it is not clear that real clients – not public health fanatics trying to prove a hypothesis but real, flesh-and-blood men who are buying sex for pleasure and want to get on with it – actually behave like this. The Indonesian data suggest not. A girl asks a guy to wear a condom, but he refuses. If it is about cash, she should be earning more than a girl who asks and uses, right? But she's not. Girls who asked their clients to use a condom but gave in to unprotected sex were paid an average of 150,000 rupiah a shot, about 16 dollars. Girls who stuck to their guns, proposing condoms and using them, racked up 210,000 for their last client – a 40 per cent mark-up compared with those who folded. I don't know why. Maybe better-paid prostitutes feel they have more to lose if they get infected with an STI or HIV.

We've agreed prostitutes have the incentive, and their bargaining power seems to be higher than we thought. So what else stands in the way of safe sex? I think about myself. I'm as aware about HIV as anyone on the planet. Yet my own history of condom use is somewhat chequered. There's the pleasure thing. There's the 'he seems like a nice boy' thing. And there's the second bottle of wine thing. I went back to the Indonesian data and checked for a link between alcohol and unprotected sex. Sure enough, women who report getting drunk with clients are a lot more likely to agree not to use a condom. In fact, they are a lot less likely to ask their client to use a condom in the first place.

It is fashionable to treat prostitutes as victims, constantly subject to the threat of violence, powerless before their clients, pimps and club owners. And in some places it is very largely true. That's why working with the Powers That Be, blackmailing the people at the top of the food chain so that they enforce condom use lower down, still has the potential to work in many settings. But women are using condoms fifteen times more often just by telling clients to use a condom; hardly powerless. And programmes that have worked with sex workers to increase their bargaining power and negotiation skills have worked well in some settings, especially where turnover in the workforce is low. But let's be realistic about women who sell sex. There is no reason to assume that they are any more or less virtuous or diligent than women who sell software, or legal services, or vegetables. The data suggest that when sex workers don't use condoms, it is often because they are too drunk to care. If we had data on software vendors, it would probably tell the same story.

Traffic jam

The 'victim' thing takes us back to the religious convictions of right-wing voters in the United States. In recent years they have launched a crusade to equate prostitution with human trafficking, and they've scored some important converts.[31] 'Sex trafficking' has assumed a life of its own, even appearing as an 'affected population' on the UNAIDS website.[32]

Some people are indeed kidnapped and sold into prostitution, and some are more or less treated as slaves. Their stories, told in any number of dramatic television documentaries, are appalling.[33] The UNAIDS web page on trafficking claims that 'Every year, millions of women and children are bought and sold . . . into the sex industry and lives of abject exploitation.' Millions. Every year. There is no source for this figure, but let's do some rough sums. Over 60 per cent of the world's population lives in Asia, and since patterns of sexual networking support a large sex industry, the continent is probably home to something like 75 per cent of the world's sex workers. It is estimated that around 12 million women sell sex in the region.[34] The average time in prostitution varies from country to country, but is generally between three and four years, so on average not more than a quarter of women take up the trade in any given year. That's 3 million new sex workers each year in Asia, and by extension, a maximum of 4 million globally.

If 'millions' of women are trafficked into prostitution every year, then a high proportion of all sex workers must be trafficked. I've been hanging around the sex trade in Asia for a good few years. I must have chatted to hundreds of sex professionals in that time, from the US$500 a night cell phone call girl to the illiterate teenager in the sooty shack behind the dockyard, and I've combed through data from tens of thousands more. Most women I've spoken to say prostitution is not their dream job, but I guess most women would say that about flipping burgers in McDonald's too. A lot of women report being introduced to a broker for brothels by a family member or a girl from the village who had trodden the path to the brothel before them. Some of these women were lied to, were told they were destined for other jobs and genuinely had no idea that they would be forced into selling sex.

I know I'm least likely to meet the women who really are held under lock and key against their will. But still it seems to me that the 'millions' are remarkably well hidden. I have only ever met one girl who said she was trafficked.[35] She was a Vietnamese girl who I met in a temple in Cambodia in 1996. When I first saw her she was sandwiched between two minders, being forced forward on her knees to bow and scrape at the saffron hemline of a Buddhist monk. She proffered a slip from the Pasteur Institute bearing the results of her positive HIV test; the monk glanced at it, tossed it aside and gave her nutritional advice: don't eat fish, meat or eggs. Then the charlatan sold the minders a home-made cure for AIDS, promising that the prostitute in whom they had invested would be hale, hearty and HIV-free within six months.

I never knew the girl's name, but she told me (while her minders paid the monk) that she had been sold by an aunt while her mother was ill and bundled across the border into the hands of these two witches. I am sure her story would have made a good documentary. But is trafficking really the norm in the sex industry in Asia?

It certainly was in East Timor, if you believed the rumours swirling around the country when I was doing some work there in 2003. East Timor used to be part of Indonesia, but became independent in 2001 after an extremely bruising conflict which left the tiny country in tatters. The United Nations, which had failed to put a lid on the conflict, hung around to help rebuild the place. Lots of young soldiers a long way from home in a fiercely Catholic country: any trafficker worth their salt should be able to spot a market opportunity in that. And sure enough, the stories of people-smuggling popped up. 'A former UN officer alleges that women are being trafficked from South-East Asian countries into East Timor to work as sex slaves and that some UN officers are customers at brothels where these often under-age women are made to work,' the Australian Broadcasting Corporation reported in 2003.[36]

I was working with the East Timorese health ministry to plan a study of HIV and risk behaviour in the nation's capital, Dili, and I'd visited every brothel in town. I'd been to a couple of massage parlours where Thai women served clients, most of them UN staff, and I'd chatted with some Vietnamese girls who took on Asian businessmen. The girls were upset that they couldn't use roaming cell phone services to send text messages home, complained about the personal hygiene of their clients and marvelled at the good pay – US$50 for a 'short time' was the standard rate. But no one said anything about trafficking or under-age colleagues. I asked Tim Hudner, a friend who worked in the political division of the UN. He said his office was thick with rumours, and promised to try and get the inside scoop from the UN cops. But by the time I went back to Jakarta he hadn't been able to unearth any real evidence of trafficking.

A couple of months later, I was back in Dili. The day after I arrived, I spoke to Tim. 'Oh, are you still interested in trafficking?' he asked, almost casually. 'I've had a report of three Chinese girls trafficked in. They say the trafficker was a UN staffer.' Good story! I was a bit surprised they were Chinese – in this part of the world 'trafficking' is a word more often associated with Vietnamese or Javanese women. But there may be a growing taste for Chinese women. Those three Chinese hookers I was chatting to at the airport yesterday, for example . . .

Hang on a minute. When did this happen, Tim? 'Just yesterday.' And the UN staffer, who was that? 'Some white woman. We haven't been able to identify her yet. She was taking care of all of their paperwork.'

So there it was. The girls had been on the same flight as me. From their clothes, their hair, their make-up, I'd guessed they were on the game. Never one to skip an opportunity to keep up with the industry, I sidled up to them in the chaotic immigration queue and dusted off my Chinese small-talk. I found out that a local businessman was offering girls three-month stints selling sex to the Chinese community in Dili. 'We were really lucky to get in,' said one girl. They got the nod from a friend who had done the run six months earlier. 'She bought a car when she got home.' Admiration all round. I helped them fill in their forms and promised to visit them soon at the restaurant where they would be based.

Now I was a sex trafficker.

The fact is, most women sell sex for the same reason that people flip burgers in McDonald's, clean other people's toilets, hack coal out of a mine or do any number of other poorly paid, unpleasant and sometimes dangerous jobs. To make money.

Sometimes quite good money. A sex worker usually earns quite a bit more than a woman making sports shoes or jogging outfits in a factory. Granted, it is not nice work. You sit around in dark, sweaty, airless rooms with the doof-doof background noise thudding in your ears, being leered over by some guy who has more power than you. That's in the factory. In the brothel, it's more or less the same, except that you take your clothes off and allow the leering to go further. The other difference, of course, is the pay packet. In the factory you earn 19 cents an hour. In the brothel your take-home pay averages about US$3.15 an hour. Two horrid jobs; one pays sixteen times more than the other.[37]

I don't doubt that some pimps and brothel owners hold women and young girls against their will, forcing them to sell sex and sometimes even keeping all of the payment for themselves. But these cases of slavery appear to me to be relatively rare. For many women, selling sex is a job with a fair degree of freedom and for some there's job satisfaction, too. Many clients want far more than just a quick orgasm. They want companionship, advice on how to cope with girl trouble, pampering to help them forget a lover's death or a business deal gone wrong. They want their confidence boosted or their scars healed, they want to learn new tricks in bed or they just want a massage and a cuddle. A skilful sex worker will read and fulfil her client's needs, and many will be well rewarded for doing so.

In Vietnam, sex workers reported earning seven times more than the general population in their area. In Nepal, six times. In Cambodia, five times.[38] This is money that goes into their own pockets – the pimps or brothel owners have already taken their cut. These averages obscure huge variations in earnings. But it isn't safe to assume that the money is bad even at the bottom end of the market.

When I was trying to figure out the sex industry in the small Chinese border town of Dongxing, I chatted with the Bai minority women on the bottom rung of the prostitution ladder. They sit in clusters in unpaved side streets, gossiping, knitting, waiting for clients. With no electricity, it's hard for them to set out the red lanterns that signal their trade, so they've improvised by trapping a candle inside a red plastic toilet-roll holder that glows a lurid pink. They registered no great surprise when I wandered over, cracking out the melon seeds and sending to the hole-in-the-wall shop across the road for tepid beer as they would for any client. The shop owner keeps an eye on the girls, and there does seem to be some further organized supervision. As we sat chatting and fiddling with melon seeds close to midnight one evening, a guy roared up on a monstrous Harley Davidson knock-off. He had the leathers, the biker boots, the bandana – the uniform of men-on-monstrously-large-motorbikes around the world. But there was a local touch. Around his neck the biker wore a jade pendant. Not a discreet little amulet but an elaborately carved disc of precious rock the size of a DVD. He had a friendly chat with the girls, was satisfied that all was well, and roared off. No money changed hands. When I asked who he was, they waved my question away with a vague 'our friend'.

It seemed likely to me that the biker was their pimp. It seemed phenomenally unlikely that he was financing even his petrol consumption on his cut of their earnings. At this rock-bottom end of the market, the girls must scrabble for every penny, I thought. I was disabused by Meiling, a girl in her late teens who had been on the game about a year. While we were chatting, an in-your-face luxury SUV pulled up a few metres away. There were three guys in it, and Meiling and a friend wandered off to talk tricks. Minutes later they were back at the melon seeds. 'No deal?' I asked. 'They only wanted to pay a hundred each.' She looked grumpy. That's 100 renminbi, about US$11 for half an hour's work. Nearly three times what you would get paid for half an hour flipping burgers in McDonald's in New York, let alone in Beijing.[39] I observed that it didn't seem like bad money. 'Not good enough for me,' Meiling said. 'And anyway' – more grump – 'they were only driving a Pajero.' In the time we sat chatting, four lots of clients were turned away, and only two deals were done (one with men in a BMW with plates from China's gambling epicentre of Macao). The 'all sex workers are trafficked' ideologues may damn me for saying so, but these ethnic minority women, working at the bottom end of the sex trade in one of the poorer areas in China, did not seem to be driven by desperation.

The hypocrytic oath

Lots of women (and men and transgenders too) sell sex because it is the best gig they can get. Even in Burma, one of the twelve countries accused by the US State Department as the world's worst offenders in people trafficking, World Health Organization researchers found that 'Most sex workers have chosen this occupation, as part of their limited livelihood options.'[40]

Conservative Christians in the United States find this impossible to accept. 'I doubt that there are many who enter the sex trade willingly, but those termed "volunteers" have been victimized in a different way . . . It is all simply evil having its way upon the soul, bashing away at the fragile imagio Deo that evil cannot abide in the human breast,' writes Garry Haugen, president of the International Justice Mission, an NGO that would like to abolish the sex trade. Brenda Zurita, a bedfellow in the accommodating embrace of US Christian organizations, says, 'Prostitution and sexual trafficking are inextricably linked and abolition is the only answer to end the horrors of both.'[41]

Abolition: of something that has existed for as long as there has been commerce; of something that provides a living for millions of people around the globe – a better living than capitalism and free trade have yet been able to offer. I sometimes wonder if Brenda Zurita and her ilk would be willing to pay sixteen times as much for their sports shoes – enough to put a hot, sweaty, miserable factory job on a par with a hot, sweaty, miserable job selling sex.

Could we abolish prostitution in the way that we have more or less abolished slavery? To wipe out prostitution, you'd have to wipe out the poverty that fuels the supply side. A noble goal. But the data we saw from young men in Thailand suggest that prostitution is driven by demand more than by supply. So you'd also have to wipe out whatever it is that makes men buy sex.

The US government is committed to doing this. It is right there in the PEPFAR plan itself: 'The Emergency Plan will support interventions to eradicate prostitution.'

PEPFAR's efforts to eradicate prostitution were carried out under the watchful eye of the programme's first director, Randall Tobias, who went on to lead USAID. In April 2007 Tobias quit his job. His number had been found in Deborah Palfrey's address book, and Deborah Palfrey, aka the DC Madam, is under investigation for running a 'prostitution ring'. Tobias (Randy, to his friends) said he did hire some of Palfrey's employees, not for sex, you understand, but for a massage. He paid them something like US$300 an hour just to rub his tired muscles. Now if there were more jobs like *that* available to women . . .

Tobias clearly understands the challenges involved in wiping out the sex trade. Perhaps to stiffen the resolve of those facing these challenges, Congress came up with the 'loyalty oath'. Anyone taking any money from the US government to do AIDS work has to sign a statement saying that they oppose the practice of prostitution, even if they have no intention of using any American money for prevention programmes related to sex work.[42] In the field of AIDS, no other country to my knowledge forces its ideological-theological-political agenda down other people's throats in quite the same way. Some have choked on it. Brazil, for example, believes it will help more people by regulating prostitution than by trying to abolish it. The Brazilian government trod on a glorious number of toes while it was forging one of the more successful national counter-attacks against AIDS. The government refused to sign the loyalty oath against prostitution, and promptly lost US$40 million in funding from the United States.

At least two US-based organizations have sued the Bush government, protesting that the loyalty oath bulldozes their right to free speech. They accept that the government can set up ideological hurdles for its own money. But they don't think the US government should be allowed to dictate what they do with money that they get from any other source.*

The loyalty oath is based on the belief, no, the absolute conviction, that anything that improves work conditions for prostitutes serves only to bind them into slavery. The High Priestess of this view is a US academic named Donna Hughes, who pontificates on the evils of commercial sex from every available pulpit. In an op-ed piece titled 'Aiding and Abetting the Slave Trade', she railed at a programme that taught Cambodian

* Condom promoter DKT and the Open Society Institute, which works with prostitutes in Eastern Europe and elsewhere, both won their suits against the US government. The administration appealed. The decision in favour of DKT was overturned in April 2007. At the time of writing, the OSI case was unresolved. But in July 2007, USAID issued guidance which seemed to create a fraction more elbow-room for US organizations wanting to work with sex workers. See www.usaid.gov/business/business_opportunities/cib/pdf/aapd05_04_amendment1.pdf.

sex workers to negotiate condom use with their clients. The programme was part of a national effort that sent new HIV infections in Cambodia crashing to fewer than 6,000 a year by 2005, from over 42,000 a decade earlier. But it was wicked. 'The Bush administration needs to . . . shut down unethical "interventions" with women and girls in brothels. Those who lack the moral capacity to know that slaves need freedom should never get funding again,' preached Hughes.*[43]

So the solution to the HIV problem in Asia is not condoms but missions to free slaves. There are plenty of willing missionaries. The International Justice Mission (nickname: Cops for Christ) flies a team of lawyers and former policemen into a country, visits brothels and documents cases of sex slavery, focusing especially on children in prostitution. Within a few days they stage a raid, dump anyone they've rescued on some local organization and fly out again. Their documentation is rigorous and their flair for publicity remarkable.[44] They deserve credit for contributing to a crackdown on child prostitution in Cambodia, where they have recently opened an office. IJM's website carries the tag line 'Our prayer is that God will be honored in all we do'. Honouring God is all well and good, but IJM would be better liked in the countries where they stage their raids if they also honoured the country's laws, and if they were a little less cavalier about the fate of the women they seize.

* The following year, I was chairing the Epidemiology and HIV Prevention track of the biennial international AIDS conference, held that year in Bangkok under the warming theme 'Access for All'. Thinking that these conferences are about putting your arsenal of facts on the table and may the better science win, I suggested inviting Donna Hughes to debate the merits of abolition versus safer working conditions for prostitutes. I was immediately shouted down by sex worker organizations. They felt that it would be counterproductive to debate with someone they believe puts ideology before facts. In the end, we did not invite Hughes to speak. So much for access for all.

That, at least, is the opinion of Heng,[45] a moon-faced Cambodian policeman whom I met over lunch at a friend's house outside the capital Phnom Penh. We were sitting on the veranda eating roast frog and watching families punting along a backwater of the Mekong in a lazy, Sunday afternoon kind of way. The ashtrays were overflowing, the lunch table was crowded with the corpses of empty beer cans, and one of the party had just sloped off to bring out another case. All was golden, all was well with the world.

Then I mentioned the NGOs working on sex trafficking.

'Those fucking NGOs,' Heng exploded. 'Excuse me, Madame, for the bad language but they should fucking *fuck off*!' He was so cross that he had trouble enumerating the iniquities of the anti-trafficking and abolitionist groups. But the list of sins included rounding up people at random, arresting people and holding them against their will and bribing girls, pimps and cops to give evidence. 'And for this bullshit, the donors give them millions of dollars,' Heng smouldered. 'The Cambodian police do all the work, and the NGOs get all the millions of dollars. The *shits*! Excuse me, Madame, for the bad language.'

In Heng's eyes, the fig-leaf of cooperation with local authorities claimed by foreign-funded anti-trafficking NGOs is not large enough to cover their naked greed.

NGOs say they work closely with local police. But the cops are not exactly squeaky clean. 'You have to be really careful who you are dealing with,' said Emmanuel Dialma, who manages another anti-trafficking organization in Cambodia, AFESIP (Acting for Women in Distressing Situations). The NGOs often keep their planned raids secret from the police until the last minute, which makes the police think they are self-serving, untrustworthy and trying to bypass national laws. So the police keep mum when they get information about forced prostitution, which makes the NGOs think the police are self-serving, untrustworthy and trying to bypass national laws.

What happens to women and girls who have been rescued? If they have been trafficked across a border, they are sometimes schlepped back to where they came from and dumped with the police there. In Northern Thailand, Burmese women are reported to have used sheet-ladders and climbed out of windows to escape their post-rescue 'shelters' and likely deportation.[46] In less dramatic cases, rescued prostitutes are handed over to a local NGO like AFESIP. There, they get to enrol in a six-month skills training programme, sewing mostly. They don't get paid.

'Look, if I could afford to go to school for six months, I wouldn't be selling sex,' says Keo Tha, a prostitute who is also shepherd to the Women's Network for Unity, an unofficial trade union for sex workers. Today, she's hanging out in her daytime office, a boat on the Mekong River shared with other NGOs. At night, you're more likely to find Keo Tha and her colleagues on the next-door boat, the one with screaming neon beer signs and twangy music, the one where guys come to dance and drink and find a partner who will stroke their egos and other bits for an hour or two.

Keo Tha and friends are discussing whether they want to be part of a drug trial being proposed by California-based Gilead Sciences. She takes time out from her meeting to tell me what she thinks of rescue programmes. 'For people like me, I'm thinking about my immediate problems, about my family. I spend months getting trained, and there's no money. At the end, if I even get a job, its 100 riel to sew a T-shirt, 300 riel for a pair of trousers.' Cambodia's local currency, the riel, is only used as small change. Even by local standards, Keo Tha's sewing income was very small change indeed – less than three US cents per T-shirt, eights cents for trousers. 'On a good day from selling sex I make five dollars.' She'd have to make 167 T-shirts a day to top that.

The rescuers are not entirely surprised that many women reject the better, prostitution-free life they would like to offer. 'The thing about selling sex is that the prostitute can feel the money in her hand at the end of each day,' said AFESEP director Emmanuel. The trouble is, not many days will be the sort of five-dollar 'good day' that Keo Tha was talking about, according to Emmanuel. He was a handsome young lawyer in a flawless linen shirt, only recently arrived in Phnom Penh from France. There was not a trace of the 'prostitution equals trafficking' ideologue about him, but he does believe that most girls would rather get off the game if they could. 'It's a shit life, in shit circumstances,' he shrugged.

AFESEP says it can't compete with prostitution's daily cash flow during the skills training period, so about 70 per cent of rescued girls who are brought to the centre check out again as soon as they are allowed to. The organization provides lots of support for the women that stay the course. It helps them identify business opportunities, gives them tools and materials to get set up if they need them. After they leave the centre and get started with their new skills, they are visited by health staff and a psychologist once every three months. And still, Emmanuel says, another two out of every five drift back into prostitution. He estimates that his organization, the largest of its kind in Cambodia, helped around 140 women or young girls get out of prostitution, and stay out of it, each year between 2000 and 2005. That's considerably more than fly-in-fly-out groups like IJM, whose well-publicized raid on a brothel area outside Phnom Penh netted thirty-seven girls, not counting those that bolted back to their brothels.[47] In 2005, IJM also netted

US$5 million from the Bill and Melinda Gates Foundation to fight sex trafficking and the HIV that it is said to fuel.[48]

By the end of 2005, fewer than 1,000 women had ever been successfully rescued from prostitution in Cambodia. A nationwide campaign to make sure guys used condoms when they bought sex had saved an estimated 970,000 Cambodians from HIV infections by the same date.[49] Yet freeing slaves is 'moral' and promoting condoms is 'unethical'.

I went to Sunday School as a child, and I still go to church every now and then. But I am completely unable to understand religious convictions or moral ideologies that stand in the way of saving hundreds of thousands of lives.

7

HIV Shoots Up

I believe that treating all sex workers as though they are the helpless victims of trafficking is short-sighted and counter-productive. But it does have a potential upside. If prostitutes are trafficked, then they are 'innocent victims' rather than 'wicked people'. Donna Hughes may think it unethical to provide innocent victims with information and services that could save their lives, but others would disagree. Voters don't exactly jump at doing nice things for sex workers, but it's certainly an easier sell than doing nice things for drug injectors.

Nobody is painting drug injectors as exploited or victimized. If you listen to the rhetoric of most Western governments, we're at war with drugs; when drug users become cannon fodder, few tears are shed. A lot of more or less law-abiding taxpayers, even ones who regularly relax with a spliff or fuel their parties with cocaine, don't want to pay to help junkies inject more safely. People who inject themselves with damaging, addictive and illegal substances do it willingly, even avidly, the law-abiding dope-smokers reason. Let them get what's coming to them.

In any war there must be two sides, and the war on drugs is

no exception. Each side holds to its beliefs with a fanaticism that is religious in its intensity, if not its origin. As is often the case in a war, they agree on what the problem is. But they disagree violently about the solution. On one side are the Drug Warriors. On the other are the Harm Reductionists.

Here's what they agree on. Drugs are harmful. They can screw up your body, your life and the lives of people around you. Because they are illegal and often expensive, they can lead to an increase in crime. Injecting drugs can also infect you with unpleasant and often fatal viruses, including various strains of hepatitis, and HIV.

Then the views diverge. The Harm Reductionists recognize all of the above. They also recognize that drugs can be a lot of fun. That's why there will always be a market for drugs, even when taking them eventually leads to lying to your boss or lover, stealing from your mother or neighbour, getting infected with a virus that will kill you. Harm reduction programmes often try to help people quit drugs, but that's not their main focus. They believe that most people grow out of drug use eventually. So they focus on providing services that will keep people alive, socially stable and disease-free while they are still using. They give people drugs they can swallow (like methadone), in the hope that they will stop injecting. And they give people who do shoot up sterile injecting equipment, so that they don't pass around needles and diseases. In other words, the Harm Reductionists think injecting drugs is harmful. But they believe the harm can be reduced by providing good services to injectors.

The Drug Warriors see the whole edifice of harm reduction as a sham, built to cover up the true agenda: legalization of drugs. Needle distribution is the toxic core of the edifice. Making sure drug injectors can get clean injecting equipment is tantamount to encouraging people to use drugs, they say. And encouraging drug use increases the spread of HIV. The only solution to HIV among drug users is an all-out war on drugs, one that sprays poppy fields from the air and get armies to chase down traffickers flying through misty mountains in rickety planes. One that punishes growers, processors and dealers. One that imprisons users and denies them access to simple things that would keep them alive for long enough to get clean.[1]

All wars have their propaganda machines. I get assaulted by e-mails from both sides in the drug war. The Asian Harm Reduction Network sends me e-mails that claim that clean needles are the answer to life, the universe and everything. My father sends me e-mails that claim that needles are the devil's work.

My father Roger is an unlikely conduit for rants against the evils of addiction. He is passionate about Irish whiskey, not-quite-Cuban cigars and the New York Yankees – all of them addictions that can destroy your health and/or your sanity. He's got no history as a proselytizer – he's spent his whole career in advertising, selling toothpaste and washing-up liquid. And yet there he was, enlisted in the war on drugs as a lieutenant at the Partnership for a Drug Free America. The Partnership brings together advertising agencies and the media to forge campaigns that discourage drug use.* In the depths of a salsa bar in some basement in Paris in the small hours, we had the father–daughter 'what the hell are you doing with your life?' conversation. Except that it was daughter asking father. Roger swirled his whiskey around in his glass reflectively, and disagreed that he had no history as a proselytizer. He'd been

* The most famous was probably a fried egg sizzling in a pan, captioned 'This is your brain on drugs'. I later gave my father a T-shirt with a picture of a slap-up fried breakfast and the slogan 'This is your brain on drugs with bacon and a side order of toast', but he never wore it.

preaching the virtues of a particular brand of useful but boring products all his life. Now he couldn't resist the challenge of using those same proselytizing skills to unsell products that were stupid but fun. Look at what was happening to smokers, he said. Not here – he waved his cigar dismissively through the haze of Gauloise fumes that in those days still settled over all Parisian bars – but in the States. We laughed at the memory of the two biddies who harangued him for smoking at the beach near New York. He was out rollerblading at some speed against a bracing wind at the time. I would have thought that was a pretty good indication of overall well-being in a seventy-something-year-old, but it was not enough to overcome the stigma that is now associated with tobacco use.

My father would like to make illegal drugs as uncool as cigarettes are becoming among young people. Both the Warriors and the Reductionists support what they call 'demand reduction'. The fewer people that take health-threatening drugs the better. But what to do about the ones who insist on getting high? Roger is no longer at the Partnership, but he still helps Latin American countries set up schemes on the same model. And he still gets informative e-mails from a different anti-drug activist group with a confusingly similar name, the Drug Free America Foundation, which he sometimes forwards to me. They describe studies that prove that syringe and needle exchange programmes spread viruses and are associated with failed antiretroviral treatment and early death.* I reciprocate by sending him some of the e-mails I get from the Asian Harm Reduction Network. These

* Needle exchanges give out clean injecting equipment, usually including syringes, and take used equipment out of circulation. I use 'needle exchange' as a generic term for any programme that provides easy access to sterile syringes and needles, whether or not it requires exchange. The currently fashionable term is the clumsier NSP or Needle and Syringe Programmes.

describe studies proving that needle exchange programmes reduce the spread of viruses and are associated with more access to antiretroviral drugs and better health.

Often, they are the same studies. It's a matter of how you read the evidence.

Here's the evidence. Scotland made it hard for injectors to get needles throughout the early 1980s, and that made it easy for injectors to get HIV. The country woke up to this in 1985, when drug injectors started turning up in Edinburgh's hospitals and in the pages of its newspapers, dying of AIDS. A year later, Scotland became the first country officially to begin distributing needles to injectors. It was too late for Edinburgh, where half of the city's injectors were already infected. It took seven years' worth of doling out needles to bring infection rates down from 55 per cent to 20 per cent. Glasgow, just 40 miles down the road, had a miraculous escape. The virus hadn't yet wormed its way into the fraternity of injectors when needles became available. Because needle distribution exploded before the virus did in Glasgow, HIV prevalence never went over 2 per cent among injectors in that city.[2] Australia, England and the Netherlands all followed Glasgow's example, making clean needles available to anyone who wanted them before HIV became well established among injectors. In all three countries, HIV among drug injectors stayed low. The pattern was confirmed around the globe. In the 1990s, HIV among injectors *fell* by 19 per cent a year in thirty-six cities with big needle exchange programmes. In sixty-seven cities without programmes, HIV *rose* by 8 per cent.[3]

When in 1992 New York City blew a raspberry at the disapproving federal government and opened a needle and syringe exchange programme, half of the city's injectors were already infected with HIV, and among those that weren't, one in twenty-five was becoming infected each year. After ten years supplying clean needles, new infections had been slashed by three-quarters.[4] And it's not only in rich countries that sterile needle programmes work. In China's Guangxi province, the sharing of dirty needles has halved since drug users could get clean ones at subsidized prices from pharmacies, no questions asked. In Bangladesh, Brazil, Iran, in city after city and country after country, people who can easily get clean needles share less than those who can't.[5]

'Aha!' cry the Drug Warriors. 'But what about Vancouver?' Vancouver is one of a handful of North American cities, most of them in Canada, which started large needle exchange programmes in the late 1980s. At that stage, it didn't go as far as Switzerland and Australia, which provide rooms where people can come and shoot up their drugs in a safe environment, watched over by staff who know what to do at the first signs of overdose. But it did give injectors a place where they could get sterile needles from friendly staff who would also hook them up with drug treatment programmes and social services. The programme was hugely popular; by 1993, Vancouver's needle exchange was passing out a million clean needles a year and rising. HIV stayed lowish for a while, so the government was happy with the programme too. Then in the mid-1990s HIV exploded. And infection rates were highest among people who used the needle exchange most frequently.[6]

This was seized upon by the war-on-drugs crowd as evidence that needle exchanges don't work. Needle exchanges just provide an opportunity for a lot of junkies to get together in the same place, meet new injecting partners and infect one another.

My friend Adrian had a different view. Adrian is a yoga teacher, and he's about as clean-living as anyone I know. He's up at dawn daily, he meditates, he's a vegetarian, naturally. All of this has contributed to a distractingly perfect physique that he swings absent-mindedly into glorious contortions. I hang around after class to chat, like a good yoga groupie. We start talking about my work, about HIV, condoms and clean needles. 'I don't get it,' says Adrian. 'Why don't people just use clean needles?' I explain that needles aren't always available, that in some places people get arrested just for carrying them, that not everywhere has a needle exchange. I'm about to start explaining what a needle exchange is when Adrian pulls a face. 'When I was using heroin I didn't go near the needle exchange. Those places are for junkies. They're for people who are just about to die.'

I was astounded when delicious, clean-living Adrian popped out of the closet where former heroin users hide out. But I was surprised, too, at how neatly he had hit on the very thing that the Vancouver researchers thought might explain their findings. They had been taken aback at the wave of *Schadenfreude* triggered by their paper, and they wanted to gainsay the people who read it as 'Needle exchanges cause HIV'.

We like to think of scientists as completely neutral, but no human is truly neutral. Most scientists choose a field of research because they have an inherent interest in it – gay men are vastly over-represented among epidemiologists working on HIV, for example. That inherent interest is going to affect the questions they ask and the hypotheses they form. This may not apply to people decoding DNA in a lab, but most science and almost all field epidemiology happens in the real world. If you dedicate your life to doing underfunded, difficult and sometimes dangerous research on the effectiveness of needle exchange programmes, you are probably predisposed to think that needle exchanges are a good thing. If you are a good scientist you won't mess with the results of your research, but you will do your damnedest to explain the bits that don't come out the way you'd like.

As I learned in my very first epidemiology lecture, the first thing to do is to blame the study design. You may remember that a cross-sectional study doesn't tell you anything about cause and effect – are you depressed because you're not getting laid, or are you not getting laid because you're depressed? People who used the needle exchange a lot had higher HIV infection rates than people who didn't. So needle exchanges give you HIV, right? That was the Drug Warriors' reading of the data. The Harm Reductionists reached for the other obvious explanation: having HIV makes you use the needle exchange. Drug injectors with HIV may have other health concerns, they reasoned, a more spotty employment record, a less organized life and may therefore be more in need of a public needle exchange service than an injector who is a doctor or an airline pilot. Adrian's observation – needle exchanges are for people who are about to die – would support that.

What does a good epidemiologist do in a situation like this? Design a cohort study that follows only the *uninfected* people over time. If we think needle exchanges protect against infection, we'd expect to find that those who use the programme a lot stay safe, while those who don't get infected.

The studies were duly designed. And as expected, they found . . .

Oh, bollocks! They found that uninfected people who used the needle exchange a lot were *more* likely to get HIV.

The researchers pored over the data, trying to figure out what was going on. What they found was that drug fashions were changing in Vancouver. Injectors started switching from heroin to cocaine. Smack injectors tend to shoot up two or three times a day, but their injections are regular, predictable and can be planned for. The coke high is shorter, so you have to inject more often to stay happy. More importantly, coke injection is a lot less predictable – injectors often go on binges injecting many times over a short period. Binges are harder to plan for, and when someone's on a cocaine roll they don't care about anything except the next hit. If they run out of needles before they run out of drugs, they're not going to pop down to the exchange for more. And in any case, they couldn't get more. The exchange in Vancouver had calculated injectors' needs based on heroin injecting patterns and were rationing needles at that level.[7] Where injection frequency goes up but access to needles doesn't, needles will get shared and HIV will go up, too. 'Aha!' cry the Harm Reductionists. 'Needle exchanges *do* work. We just need more needles.'

Adrian's comment got me thinking. 'Needle exchanges are for junkies!' he'd said. Yes, he used smack, but he wasn't a *junkie*. Blerch. Junkies are gross, they're disorganized, they cheat and lie and steal, they've let drugs become their life. For people like Adrian, drugs are just one thing that goes on in life, along with a job and a girlfriend and hitchhiking around Ethiopia, going home for Dad's sixtieth birthday, hanging out at Starbucks. When I started to look for information about these people I found that only three in ten heroin users are 'junkies', shorthand for drug-dependent. The other seven are people like Adrian, people who will use for a little while and then clean up. This proportion comes from US data. In a massive national study in Canada, one adult in thirteen said they had used injectable drugs at some point.[8] Not all of them were active injectors. But apart from the 0.3 per cent who used illegal steroids, they had all used hard drugs that are usually associated with injecting, with being a 'junkie'. Look around your office, your lecture hall, your church. Count off every thirteenth person. Think there are that many junkies around? No. There's anecdotal evidence that a disproportionate number of doctors inject prescription drugs illegally. The others are yoga teachers or stockbrokers or taxi drivers or pilots. These people are not going to use a public needle exchange programme on the corner of a blighted inner city block, but they do need a legal framework that allows them to get clean needles easily. We know virtually nothing about what else functional drug injectors might need, because so much of the research focuses on the bottom-rung addicts, the people who inject in toilets on station platforms, the people who bounce in and out of rehab, the people who end up in jail. These are the people who need publicly funded HIV prevention services most of all. But they are also the people who are most of a turn-off for politicians and voters.

The occasional users are actually a real worry, especially where HIV prevalence is very high. Chris Green, who runs HIV support groups in Jakarta with equal measures of passion and common sense, talks of people with HIV who don't seem to be at risk. They're not gay, they don't buy or sell sex, they don't sleep around much, they don't inject drugs. With a bit of digging around, it may transpire that they did try heroin once, maybe twice, a couple of years back, when a bunch of their friends were shooting up. But they didn't like it and they never did it again. For HIV once or twice is enough, especially when you're shooting up with buddies, and half of them are infected.

As we saw earlier, you can have quite a lot of unprotected sex with someone who is infected with HIV and, as long as you don't tick one of the 'danger boxes' – newly infected partner, other sexually transmitted infection raging, unlubricated anal sex – you will in all likelihood get away without catching the virus. But sharing a needle with an infected injector is an automatic red alert. If you are going to shoot up drugs, you need to use your own needle and only your own needle, every time. To be effective at the level of a whole city full of injectors, needle programmes need to be huge. In 2000, the needle exchange in

Vancouver gave out 273 sterile needles per injector, and it still wasn't meeting demand. The same year, the needle exchange in Vietnam's Ho Chi Minh City gave out two needles per injector. Not two a day, which would be just about ideal. Two a year. At the start of 2000, 40 per cent of junkies in Ho Chi Minh City were infected with HIV. By 2001 the rate had doubled to 81 per cent.[9] And I do mean junkies in this case. Almost all the HIV data as well as information on injecting behaviour come from studies of addicts whose lives revolve around drugs.

Of course clean needles don't have to come from a needle exchange. As long as injectors can get clean needles easily it doesn't really matter where they come from. In Indonesia, injectors rarely report difficulty getting injecting equipment (or 'works'). They buy them from veterinary supply shops; they especially like syringes designed for chicken shots, because they are dirt cheap and nice and small, just right for a quick high. So how come 100 per cent of injectors know that sharing needles can transmit HIV, yet 85 per cent of injectors are still sharing works?

There are a number of possible reasons. One is that people know or assume they are already infected, and just don't care very much if they pass the infection on. Or they don't think they're infected, but don't care very much if they get infected. After all, HIV takes years to kill. A lot of young injectors don't think they'll live long enough to die of HIV, even if they get infected. An over-enthusiasm for injecting combined with very variable drug purity means that new injectors often misjudge the amount of toxins their body can cope with. In Indonesia, more than one injector in four reported overdosing in the past year. Those are the ones that got help and survived to tell the tale. An astounding 88 per cent said that at least one of their immediate injecting circle has died of an overdose. China's Yunnan province has a long history of opium use and most injectors there have had long years of practice sticking needles into themselves and one another. But even there, 78 per cent of heroin injectors had friends who had died of a drug overdose.*

But 'I don't care if I infect someone else' is not an answer you give to a surveillance interviewer. Nor is 'I'll probably die of an overdose; what do I care if I get HIV?' When we asked injectors why they shared needles even though they could easily buy them in veterinary shops, two-thirds of them gave a very different answer. They said it was because they didn't want to walk around carrying a syringe. Cop + random search + syringe = arrest and jail.

Girls, who are far less likely to get searched, are less worried about this. But it becomes a headache for them too. Ling Ling, an addict from a nice family who I met during her umpteenth attempt to get clean, said she always carried her own needle. 'But the blokes are such slobs, they never have a needle with them,' so she lent them hers. 'I was really careful, though, I always used first.' Then, when her works had done the rounds, she'd put them away until the next time she used 'first'. Not surprisingly, Ling Ling got infected with HIV.

We were chatting on the back porch of a rehab centre in the hills east of Jakarta. It was a hot, sticky evening, the smoke from our clove cigarettes hung heavy around our heads. The cicadas were in full voice, outsung just occasionally by the centre's pet owl. I was longing for a cold beer, but a rehab centre is not the place for a drink. Ling Ling was telling me how she passed her HIV infection on to her daughter. She was absolutely matter-of-fact about it. 'I was one of those people for whom

* This crisis is not, of course, restricted to Asia. For example, a study in Barcelona found that heroin addicts had a life expectancy of thirty-eight, almost exactly half that of other Spaniards. A third of the 1,005 heroin users who did not get through the study period alive died of an overdose (Brugal et al., 2005).

PMTCT failed.' You can tell from the way the clumsy, English-language acronym rolls off her tongue that she has spent a lot of time hanging around the AIDS mafia. Prevention of Mother To Child Transmission centres on giving the mother antiretroviral drugs late in pregnancy and around delivery. In Indonesia at the time, it would only have been available to people with good connections. It turns out Ling Ling has worked for UNAIDS, she has even run a support group for HIV-infected women. And now this beautiful girl from a very privileged background, this girl with the courage to be open about her HIV infection, this girl who's plugged into every type of prevention and treatment programme going, is telling me that she lends her needles to any slob of a boy who is not carrying his own. But only after she's stuck them into her own HIV-infected bloodstream.

I try and I try and I try to be non-judgemental, but I rolled my eyes, just for a second. I couldn't help it. A you-should-know-better kind of roll. Not kind. The look did not escape Ling Ling. She thought for a moment, and then said: 'Oh yes, I guess that's not smart. To be honest, I've never really thought about it.' I suddenly wanted something a lot stronger than beer.

Numbing it down

Ling Ling taught me something interesting. Asked to present data to the Indonesian government about HIV prevention options for drug injectors, I had been looking at studies about methadone. Methadone is a synthetic drug which mimics the effect of opiates, the poppy-based drugs among which heroin is king. It's no help if people are injecting amphetamines or cocaine. But if the drug of choice in your country is smack, you're in luck. Methadone stops the craving for heroin, stops the sweaty, trembly, stomach-churning withdrawal that addicts so fear when trying to give up smack. You can swallow it, so you don't need to inject. That cuts down on diseases and abscesses. And it is really cheap, a few pennies a dose, so most governments could afford to give it to a whole lot of people.[10] It gets better. If you switch from heroin to methadone you can get your daily dose from a clinic. No more scrabbling around for cash, no more selling sex just to afford a fix, no more frantic phone calls and furtive meetings with dealers. No more slinking past cops hoping you won't get busted for carrying drugs or your works. No more cold sweats as you wonder where your next fix is coming from. All these upsides, and just the one downside: no more high.

Ling Ling dropped out of her methadone programme after just twelve days. 'It makes you feel like a robot,' she complained. 'Makes you numb. I don't want numb, I want high.'

A lot of people in methadone programmes go on getting high on smack. But a review of programmes involving 17,700 people shows that they do inject less.[11] And they don't feel the same urgency as someone who lives in fear of withdrawal. That means they are less likely to inject when the only option is a shared needle. If you're an injector who is not infected with HIV, you'll be less likely to get infected if you're in a methadone programme than if you're not, even if you go on injecting on the side. Injectors with HIV who are on methadone are more likely to take antiretroviral medicine correctly and their health deteriorates less quickly. On top of that, they are less likely to die of an overdose, and less likely to wind up in jail.[12]

Studies show that the success of methadone programmes depends on getting the dosage right. Too low, and smackheads just go back to full-on injecting. Too high and you get that 'numb' that Ling Ling complained about, so addicts drop out and go back to full-on injecting. 'Those people don't know anything about high,' she said of the staff at the health centre which dished out her drugs. 'The doses were just all wrong.' My heart goes out to the staff of this government-owned health centre, on the edge of the grotty, overcrowded port area in north Jakarta. They are already overrun with mothers bearing kids who've got diarrhoea from the open drains that lace the area, with cases of malaria and TB and the city's wet-season prize disease, dengue fever. They are paid a princely sum of around US$50 a month to cope with this crushing workload, and now we're piling on a bunch of junkies who are coming in for methadone every day. It's hardly surprising they don't know all that much about being high. As they learn what makes junkies happy and start to get the doses right, more people will stay in the programme. Even more work to do.

If all goes well, though, methadone programmes should work out in the long run, even for health workers. More methadone and less injection should mean less HIV, and less HIV means less of a whole slew of other health problems which would also, eventually, be piled on to their workload. Indonesia is taking tiny steps towards making this a reality. China has decided methadone is so promising that it is moving by leaps and bounds. In 2004 China had eight little boutique projects handing out methadone to junkies. By the time you read this, they aim to have a minimum of 1,500 clinics in place, serving hundreds of thousands of people every year.[13]

A captive audience

One of the first methadone maintenance programmes in Indonesia was in a prison in Bali. And it makes perfect sense. Prisons in Indonesia are HIV factories.

We first became aware of this during the data coding fiasco that I talked about earlier, when we found out that the huge leap in HIV infection that we had trumpeted globally was in prisoners, not sex workers. Everyone fell about in relief that the explosion hadn't happened in Jakarta's teeming sex industry; we didn't think too hard about what it really meant.

What it meant was that one in six of the men locked up in the building directly across the street from the offices of the health ministry's AIDS programme was infected with HIV. We did nothing. By the next year, it was closer to one in four. We still did nothing.

At first we just assumed that HIV was rising in prison because HIV was rising in drug injectors, and drug injectors were going to prison. It took another error in the surveillance system to open our eyes. Indonesia's national guidelines for HIV surveillance in prisons state that once a year, health authorities should test a random sample of 400 inmates in any given jail. Here's what happened in one West Java prison.

Year one: 1 per cent infection. Year two: 21 per cent. Year three: prison director decides there's no point testing people who may have been tested before, so only newly arrived prisoners are tested. 'Only' 5 per cent are infected. Year four: prison director rapped over the knuckles and random testing restored: 21 per cent again.* Recap: one in twenty newly arrived prisoners is infected with HIV. After they've been in jail around eleven months on average, one in five is infected.

When I first arrived in Indonesia I had argued against doing HIV surveillance in jails for all sorts of reasons.† Now, though,

* The graph looks like a roller-coaster. See the gallery section of www.wisdomofwhores.com.

† Surveillance tests are supposed to be done on anonymous blood samples, but I thought it unlikely that prison authorities would follow the rules. Having swallowed the sacred cow of testing anonymity, I worried about what would happen to prisoners who were identified as infected. On top of that, I argued that we already had surveillance in drug injectors and gay guys. It was drug injection and anal sex that were risks for HIV, not being in jail per se. I was wrong.

I was really glad of the West Java data. It made it impossible to deny that people were getting infected in jail.

Where did all those new HIV infections in jail come from? The same places almost all new HIV infections come from: sex and drugs. Well, anal sex probably, though we don't have much information about that in Asian jails. But drugs definitely.[14]

The HIV factories even have organized production lines. Listen to Frankie, a former injector from Bali. In 1997 he was put away in Kerobokan prison not far from the tourist-larded beaches of Kuta. At the time, discipline in the prison was pretty lax. The guards were bored and underpaid, a bit of cash and they'd look the other way when guests brought gifts or prisoners needed some privacy. One day, a visitor brought in six grams of heroin. 'To cheer us up,' Frankie said. Party time! Frankie and twenty-six of his injector friends got together in a room, and elected one person to dole out the drugs. 'He made us all stand in line as if we were queuing at the box office in a movie theatre,' Frankie told me. Frankie was number 21 in line. 'There was one needle. The guy would inject one person, then the next.'

What were the chances that the designated injector filled the syringe up completely with bleach twice and then refilled it completely with sterile water twice between each injector, according to recommended needle cleaning guidelines?[15] Between nil and zero per cent, I guessed, but I asked anyway. Frankie laughed. 'He'd slosh some tap water through the syringe after every couple of people, and then go on to the next in line.' Frankie shudders at the memory of it. 'Three . . . four . . . five . . .' He's counting slowly now. 'And I'm thinking only fifteen more to go. There's blood everywhere, the syringe is really kind of disgusting, it's getting blunter and blunter, but all I can really think of is the drugs. Will there be any left by the time he gets to me? Please, please let there be some left.' Now, a decade later, Frankie sounds slightly incredulous about it. 'Addiction, eh? Boy, it can make you stupid!' Then as an afterthought: 'I can promise you, though, if ever there's a place you want to be on smack, it's in jail. The time passes much more easily.' I ask whether he's still in touch with any of the other guys who were in his line-up. 'A lot of them are dead now, or in hospital.' Frankie quit injecting in 2001. That same year he found out he had HIV. 'The only drugs I take these days are ARVs [antiretrovirals]. I've even given up smoking.'

In Indonesia's jails, no one has given up anything. In 2004/5 we talked to nearly 1,500 active drug injectors in Indonesia's five largest cities. Nearly 30 per cent of junkies have been in jail at some point, and one-third say they injected inside. In fact, drugs are sometimes cheaper inside prison than outside. When I last spoke to Desi, the injector who chooses heroin because it allows for a normal life, her own husband was in Cipinang jail in Jakarta, in a cell formerly occupied by President Suharto's son Tommy, convicted for ordering the murder of a judge who was investigating a corruption case. ('Nice cell,' comments Desi. 'They even retiled the bathroom.') She said that a little while earlier there had been such a heroin glut in Jakarta's jails that she and her husband had spotted an arbitrage opportunity. When she went to visit him, she started smuggling heroin from inside jail to sell to users outside.

What's not cheap in prison is a clean needle. Or any needle, for that matter. 'Needles are much harder to get in jail than drugs are,' Desi said. They're more expensive too – three dollars for a needle, compared with two-fifty for a fix of smack. The same needle goes around and around in the crowded cells, getting blunter and blunter despite the constant whittling away at it. Blunt needles draw more blood, more blood means more efficient HIV transmission.

There's another reason that prisons are such efficient HIV factories. If you are injecting drugs or having anal sex in prison, the chances are that you are doing it within a fairly restricted circle, day after day. If someone in that circle is newly infected, you will swap body fluids with them pretty soon. Right when they are most infectious.

Prison sentences in the overcrowded jails of Indonesia tend to be short. Get busted on drug charges or some kind of petty thievery, and you're probably looking at six months or a year. If people are being infected with HIV while they are in jail, there's a good chance that they'll be released while they are still highly infectious. And what's the first thing you do when you've been stuck in a small cell with a bunch of men for six months? For lots of people: have sex. For some, shoot up, then have sex. In other words, people are trooping down to somewhere like Rawa Malang, having a beer or two and paying a girl like Nana for sex. Right when they are most infectious.

HIV in jails is not just an Indonesian phenomenon. There's evidence that prison is a good place for people to learn to be injectors. One in six injectors in Thailand said they had their first fix in jail, and the stats are similar in Russia, Canada and several Western European countries.[16] In Thailand, junkies who injected in jail were seven times more likely to get infected with HIV than injectors on the outside.[17] There's clear evidence of HIV transmission within jails from the United States to Russia, plus Ukraine, Lithuania, the Czech Republic, not to mention Australia, Brazil, Iran and Scotland. In England, a quarter of adult men in prison say they have injected drugs, and 6 per cent said they injected while in prison. Needle sharing outside prison is pretty low in the UK these days, in large part because there are over 2,000 needle exchanges, handing out over 27 million needles a year. Britain's overseas development arm funds sterile needle programmes in prisons in other countries. But the Home Office, which runs the prison service, is more conservative than the rest of the government. Unlike the prison service in Iran, the Kyrgyz Republic and ten European countries, the UK's prison system can't quite bring itself to allow inmates to swap used needles for clean ones. One result is that the government is being sued in the European Court of Human Rights by former prisoner John Shelley, who argues that prisoners are entitled to the same services as free citizens. That's inconvenient for the government. Another result of not being able to get needles in prisons is that three-quarters of injectors in UK prisons pass round their works. That is more than inconvenient for the injectors who will get infected with HIV and for their families and sex partners. And it's not all that great for the British taxpayer, who will pay for expensive antiretroviral treatment for the people who do get infected while they are in prison.

Restricted circles of young men (lots of them addicts) stuck in a small space with relatively easy access to drugs, limited access to needles, no access to women, no condoms or lubricant and nothing to do all day. An HIV-prevention nightmare. Or just perhaps, a fantastic opportunity.

HIV prevention is like any other business: start-ups always have trouble building up a customer base. But if you're selling HIV prevention for junkies, you've got a head start: as many as a third of your potential customers are going through jail. A captive audience.

Needle exchanges work really well in jail. Needle sharing has dropped precipitously in prisons with needle distribution systems (e.g. from 71 per cent to zero in Berlin prisons) and no new

HIV infections have been recorded in those prisons.* Prisons are also an obvious place to start injectors on methadone. Giving prisoners methadone cuts heroin use, cuts injecting in jail and cuts needle sharing among those who do inject; connecting prisoners with methadone services when they get out of jail provides continued contact and may help wean people off drugs and re-establish a non-criminal life.[18] Jails are also a smart place to think about preventive treatment for tuberculosis. TB is the most common piggy-back disease in people with HIV in developing countries, and it thrives in the crowded conditions of a prison. Put HIV and crowded conditions together, and you get the perfect recipe for a TB epidemic, an epidemic which can be staved off by providing medicine to people whose TB is not yet active. At the very least, we could do more early detection and treatment for active TB.

Both methadone and preventive treatment for TB are cheap, relative to many of the things we spend HIV money on – the vast conferences, the endless workshops, the nationwide prevention programmes in groups that are never likely to take drugs or have risky sex. But they are hugely expensive relative to the amount many prison services budget for inmate health. In 2002 the head of Cipinang prison, where President Suharto's son and Desi's husband were both held, said that he had a health budget of 700 rupiah per prisoner per year. That is seven US cents per prisoner per year. Enough for a couple of aspirins, if you didn't also have to pay for a nurse to dole them out. Certainly not enough for methadone, condoms, HIV and TB testing, let alone TB drugs and antiretrovirals.

* Many countries provide bleach to prisoners so that anyone sharing needles can clean them before reuse. The effectiveness of providing bleach has not been proven. See this chapter note 26 and Dolan et al., 2003a; Stark et al., 2006; World Health Organization et al., 2007.

Did you pay taxes in 2002? If you did, you probably contributed to the US$8 million that the world funnelled into HIV programmes in Indonesia that year. But not a penny of that went into HIV prevention programmes in the country's prisons. Those HIV factories were left to operate on a health budget of seven cents per person per year.

Needle in a haystack

Methadone programmes are gradually becoming more acceptable. Even the United States, which used to think that methadone came unacceptably close to doing nice things for junkies, started funding it in its overseas AIDS programmes in March 2006, though only on a pilot basis.[19] Needle exchanges are having a harder time of it. It is extraordinarily difficult to shift people from the idea that giving out needles is tantamount to telling kids to shoot up drugs. '[Needle hand-out] programs encourage illegal drug use,' declares the Drug Free America Foundation.[20] But it gives no evidence to support this assertion. I've trawled the medical literature in search of evidence that safe injecting programmes encourage drug use. I've scoured the War-on-Drugs websites. I've asked my father. All I can find is a single descriptive study of two heroin smokers who switched from smoking to injecting in a jail with a needle programme. The rate of switching was the same in jails with no needle programmes.[21] But let's not let the data get in the way of a good story.

Most people who take drugs take drugs for fun. In fact the legal drugs – alcohol, caffeine, cigars – are pushed as sources of pleasure, satisfaction, fulfilment by my father's old advertising buddies. Smoke this cigar and you'll soon be living in a European castle with a wood-panelled library. Drink this cognac and the gorgeous brunette in the velvet ball gown will just fall into your arms. You're not allowed to advertise cocaine, of course, but if you were I bet the ad would involve some hunk with a square jaw abseiling down a glacier while yelling into his cell phone to close the deal on a multi-billion-dollar investment. Smack ads would have the evening sun coming through the clouds to cast a golden light on the shimmering but infinitely tranquil sea. Of course heroin makes people feel golden and shimmery and tranquil even without the ads – that's why people start taking it. But if you're in the 30 per cent of users who become dependent, the tranquillity doesn't tend to last. Smack eats up money, friendships, lives. Which is why some people try to get clean.

Severely addicted users seem to have better luck getting clean if they are former clients of needle exchanges, despite claims to the contrary by opponents of harm reduction. This seems to be partly because injectors who turn up every couple of days for their stash of clean needles are kept in regular touch with health workers and counsellors who can encourage them to quit, and who can refer injectors to treatment programmes when they're ready to try kicking their addiction.[22] Over 80 per cent of the needle exchange schemes in England help injectors get places on treatment programmes so that they can get themselves off drugs.[23] In many countries, the biggest obstacle to getting clean is simply that there are just not enough detox treatment programmes to go around.

All over Europe, and in Australia and some countries in Latin America and Asia too, governments just quietly get on with providing services that drug injectors need. Pharmacies and specialized needle exchange programmes take a lot of used needles out of circulation, so there's less risk that HIV-infected syringes and needles will be left lying around to stab unsuspecting kids playing in parks or back-alleys. The programmes do all sorts of other useful things, too. They vaccinate injectors against hepatitis, they provide HIV counselling and testing, they give out condoms, they teach injectors how to avoid overdosing, and how to help friends who've taken too much smack. All the evidence suggests that harm reduction programmes help people quit drugs, and increase the chances that people will not be infected with a fatal virus when they do manage to get off drugs.

I find it hard to respond to the e-mails my father passes on, because other than gloating over the findings of the early Vancouver studies, they very rarely provide any data to back up their claims. Drugs are evil and harm reduction promotes drug use and that's that. Enough American voters feel this way that the United States government has maintained a ban on federal funding for needle exchange programmes since 1988.*

One of the Warrior websites contains the following baffling assertion: 'There are no scientific-based studies that prove these [needle exchange and harm reduction] programs prevent AIDS and discourage drug use.' No scientific studies? Leave aside all the individual studies I've already cited in this chapter. Here's a list of US government agencies that have conducted comprehensive reviews of the effectiveness of needle and syringe programmes, and found that they cut transmission of fatal viruses and help drug users access other services, while doing nothing to increase drug use or drug injection.[24]

* We're not talking just about the Bush administrations. In 1998 Bill Clinton was widely expected to drop the ban on funding needle exchanges, but he chickened out at the last minute. 'I was wrong about that. I should have tried harder,' he told Larry Altman of the *New York Times* in 2002 ('Clinton urges global planning to halt HIV', 12 July 2002).

- The National Commission on AIDS (1991)
- The General Accounting Office (1993, 1998)
- The Centers for Disease Control and Prevention (1993)
- The Office of Technology Assessment of the US Congress (1995)
- The National Institutes of Health (1997)
- The American Medical Association (1997)
- The United States Surgeon General (2000)
- The Institute of Medicine (2006)
- The National Academy of Sciences (2006)

Here's a list of US government agencies that have conducted comprehensive reviews of the effectiveness of needle and syringe programmes, and found that they increase drug use, criminality or disease transmission:

•

That's right. Not a single one.

The gods in the US pantheon of scientific credibility all favour needle exchange. The politicians that pay their salaries do not. And so not one clean needle has been bought for a drug injector with money from the US federal government. Not one needle in the United States, and not one needle anywhere else in the world either.

This is a nightmare for colleagues working on AIDS in organizations such as USAID and CDC. Working to reduce HIV infection among drug injectors is not exactly a glamorous job. The majority of the people who do it are dedicated, hard-working people who really want to do what works. And that's hard enough as it is, without having to start every day by looking for ways to circumvent often damaging regulations imposed by the government you serve. It's as if the proprietor of a football team has put together a really good squad, then tied the players' legs together before sending them out to face the competition.

The restrictions have been a real headache for people like Fahdli, too. Fahdli, an Indonesian colleague in his twenties, was a former heroin addict. He more or less drifted to the head of a team of young people struggling to run a drop-in centre for drug users in Surabaya. Fahdli was not short of headaches – one of them was his colleague Henry who had so enraged me by injecting in the office. I wasn't helping, either. There we were, sitting around on the floor of the drop-in centre, hot and sticky, drinking bottled tea and trying not to get distracted by the ice-cream seller who was banging his gong in the alleyway outside. Fahdli was explaining what he and his colleagues were supposed to be doing: getting young injectors to sit together nicely in support groups, where they'd discuss safe injecting and put pressure on one another not to share needles. And all I could do was roll my eyes.

'I know, I know,' he sighed, dispirited. 'Junkies don't give a shit about support groups. They want two things: smack, and needles.' Our programme didn't provide either. A few countries do give heroin to addicts on prescription as a way of helping them stabilize their lives, but nobody was advocating that for Indonesia. Plenty of people – including at least one Indonesian police general – were advocating safe injecting programmes. But because our money came from the US taxpayer, we weren't allowed to give out clean needles either. Condoms to protect the innocent women the injectors might have sex with, yes. Bleach to clean needles and hopefully kill any virus that might be lurking there even though bleach hasn't actually been shown to be very effective in real life, yes. Alcohol swabs to disinfect skin with, so that the puncture wound doesn't get infected leaving a weeping abscess, yes. Tea, sympathy, homilies on safe injecting and support groups, yes. But the one thing that has been shown to cut HIV transmission among drug injectors, no.

The result: before the programme began, nine out of ten injectors in Surabaya regularly shared needles. After two years of hard work on the part of Fahdli and his mates, nine out of ten injectors in Surabaya regularly shared needles.[25]

They were a little bit more likely to bleach their needles than before, but that was about it. Unfortunately, that's not very helpful. Bleach murders HIV in the laboratory, but it seems to lose its power to kill when it is used by real addicts in real life. Drug injectors who say they use bleach every time they share needles have been shown to be just as likely to get infected with HIV as those who used it sloppily or never.[26] In other words, we were doing things that made almost no difference to the spread of HIV. By my estimates, the US taxpayer was spending about US$70 per junkie to make no difference, and that didn't include overheads or the salaries of expensive expat staff like me. Those same taxpayers have funded umpteen reviews showing that clean needles save not weeks of life but whole lives. Reviews that show that for the price of a few hypodermics, somebody's father, somebody's wife, somebody's yoga teacher could live for an extra three or four decades, long after they've outgrown using heroin. And after all, seven out of ten users do not construct the rest of their lives around drugs.

Is it inevitable that conservative governments dig their heels in at the thought of helping drug users stay healthy until they manage to quit? Apparently not. It was Margaret Thatcher's government that funded the first needle exchanges in Britain in 1986. And as I write, one of the most conservative governments on the planet, the hyper-religious government of Iran, is busy expanding needle programmes for injectors all over the country, including in its prisons. There are around 180,000 drug injectors in Iran – fewer than there are in New York City. Around one young adult male in 100 is an injector in Iran, a lower proportion than in Indonesia or Scotland.[27] Not an overwhelming social problem, but apparently enough to prompt the government to do something positive. Some governments look at the evidence around harm reduction and act with pragmatism and compassion. Others hide from the science behind a wall of ideology. Funnily enough, it is often those that are secure in their popular majority (such as Thatcher in 1986) or those that don't have to worry too much about voters (such as Iran and China) that can afford to be compassionate. Those that look skittishly over their shoulders at every opinion poll (Washington, Jakarta) are more likely to allow dogma to overcome reason. That suggests that it is voters, rather than governments, who lack compassion.

Sharing the faith

In democracies, governments are assumed to have the right to dispose of their taxpayers' cash until the voters tell them otherwise. If Americans don't want to buy clean needles for drug users in Indonesia, that's fine. But do they really have the right to tell other countries what to do with other people's money? The Bush administration believes so, and says so quite openly. When it put US$15 billion of its taxpayers' money on the table for HIV programmes worldwide, Washington declared that it would 'actively [work] to ensure that all resolutions and commitments agreed to in the multilateral area are compatible with our bilateral policies'.[28] For those who don't speak Development Bully, that translates as: 'We'll do our damnedest to make sure that all the international organizations toe the US party line.'

And it has certainly been working actively, beginning by undermining support for harm reduction policies for drug injectors. The administration started by turning the screws on something of a soft target, the United Nations Office on Drugs and Crime, which was never much of a fan of programmes that keep junkies alive in the first place. Its executive director, Antonio Maria Costa, felt compelled to write to US State Department official Robert Charles, assuring him that 'we neither endorse needle exchange as a solution for drug abuse, nor support public statements advocating such practices'.* A few weeks later, a senior manager at UNODC quoted the letter in an e-mail to staff, adding, 'Please ensure that references to harm reduction and needle/syringe exchange are avoided in UNODC documents, publications and statements.'[29]

The next target was UNAIDS, a rather harder nut to crack. The ever-vigilant Congressman Henry Waxman said that the United States tried to pressure UNAIDS to drop any mention of needle exchange programmes from a major policy document on HIV prevention.[30] According to a friend of mine who was close to the negotiations, they nearly succeeded. Peter Piot, head of UNAIDS, 'got his knickers in a twist & was ready to cave to US anti-injecting equipment crap', the friend wrote to me. It took the concerted efforts of the Northern European nations and Britain to keep clean needles on the UNAIDS prevention agenda. Though you'd be hard pressed to know it. Injecting drug users do not appear as an 'affected community' on the UNAIDS website. 'Indigenous peoples' do. 'Rural communities' do. Oh, and 'Sex trafficking'. But no mention of drug injectors, *the* most affected group in terms of HIV prevalence in

* Letter dated 11 November 2004. It seems the men are buddies – the 'Dear Mr. Charles' is struck out and replaced by a hand-written 'Dear Bobby'.

China, Indonesia, Pakistan, Bangladesh and Russia, and in huge swathes of India, too. In other words, the most affected group in five-and-a-half of the world's seven most populous countries. Countries that between them account for just over half of the world's population.*

Peter Piot is not ideologically opposed to programmes that help junkies stay healthy. He is a scientist, he understands the evidence, and the evidence points to clean needles. But he is also the head of an organization whose biggest single donor does not give a damn about scientific evidence.[31]

Actually, that's not fair. The Bush administration does give a damn about science. It's damned if it is going to take any notice of the bits that don't accord with the views of the voters that support it so faithfully. But it governs a country that is stuffed with smart people who do great research, people who will follow the facts wherever they lead. A lot of these people work in the scientific pantheon, at CDC and the National Institutes of Health and the National Academy of Sciences. They are often invited to share their opinions with the world. Here's the conundrum: America wants to influence the world. But these scientists, if let loose, might describe the sort of data we were looking at earlier – data which show that many AIDS prevention programmes are based not on fact but on ideologies. So the Bush administration decided it would censor scientists, stipulating where they could go and what they could say. A Bush sidekick wrote to tell the World Health Organization that American government scientists weren't allowed to provide expert advice to international organizations unless

* www.unaids.org/en/Issues/Affected_communities, accessed on 24 April 2007. When I pointed this out to Peter Piot, one of his staff sent me the url for a page that does list such policies. But the page was 'orphaned' – there were no links to it elsewhere on the website at the time.

that advice happened to parrot the positions adopted by the US administration.[32]

Some scientists worry that even without the three-line whip it will soon be hard for them to advocate anything *but* government policies. Good scientists advocate policies that are based on evidence. But to get the evidence, you need to do the studies. And those are getting harder to do. In October 2003 the National Institutes of Health started asking researchers to justify studies that had already been approved and funded. A total of 157 researchers were hauled up because their work was on a list of studies that look at sex or drugs. Who drew up the list? Not any panel of scientists, not an academic evaluation body, not even a government audit agency. The list was compiled by the Traditional Values Coalition, which describes itself as a lobbying group representing over 43,000 churches in the United States.*[33]

You can imagine the e-mails that started flying around the HIV research networks when the list was leaked. We were all trying to think of ways to describe our research without hitting the trip-wires set by groups like the Traditional Values Coalition. In this campaign of intimidation, funders understandably began to get squeamish. Our good colleagues in the US development finance body USAID had always been generous in funding worthwhile HIV prevention research (including my own). They had actively encouraged us to present the results at conferences and even asked us to remind them when

* The Coalition takes a particular interest in gay and transgender issues. Its website includes 'Ministry and Counselling Resources For Those Struggling With Same-Sex Attractions And Other Gender Identity Disorders'. It also provides movie reviews. A recent offering about a popular cartoon blockbuster: 'Parents Beware: "Shrek 2" Features Transgenderism And Crossdressing Themes'.

our work was accepted so that they could publicize the presentations. My colleagues duly sent in an impressive list of research results ahead of the 2004 AIDS conference in Bangkok. But the list of presentations highlighted by USAID before the conference was horribly anaemic. Every paper that had sex workers, drug injectors or transgenders in the title got dropped from the list – USAID couldn't risk drawing the attention of the True Ideologues to work that might save the lives of the wicked.

Tongue Thaied

We saw earlier that religious rants about condoms come in many flavours, in several continents. It's the same with drugs. Most Western European countries have decided to follow the science and provide services that injectors want and need. Some countries in other parts of the world, from China to Iran, have followed suit. But the US is very far from being the only country that has taken a 'just say no' approach to harm reduction.

In 2004 Thailand hosted the Fifteenth International AIDS Conference. Once upon a time, these conferences were about science. Nowadays they are about institutional posturing, theatrical activism and money. Lots of money. The Bangkok conference cost US$18.5 million. Nearly 19,000 people rocked up to it, scrummaging for the goodie back-packs given out by the pharmaceutical companies. Big Pharma paid handsomely to nab the best real estate, the exhibition booths in the centre of the main hall. They paid again to have their booths dressed to impress, with cappuccino bars, an indoor waterfall and larger-than-life-sized photos of gleaming Western labs and grateful African children. Conference goers could admire a fabulous selection of ball gowns by Brazilian designer Adriana Bertini, all made of condoms.* They could gawp at dancing elephants and Puppets Against AIDS. Delegates paid around US$1,000 each to attend this jamboree. In many cases the tab is picked up not by the person who gets to do the admiring and the gawping, but by taxpayers in rich countries. The conference made a profit of US$1.8 million. Forty per cent of that went to the Thai government, and the rest was funnelled back to the International AIDS Society, which organizes these conferences every two years.[34]

Thai Prime Minister Thaksin Shinawatra was in an expansive mood as he made the opening speech at the Bangkok conference. I listened from the cavernous conference hall's audiovisual command centre. My computer was hooked up to the two massive screens that dominated this airline hangar; I was waiting to put up slides for two of the later speakers. As Shinawatra droned on about Thailand's HIV prevention successes, I made an idle wager with myself: bet he doesn't mention junkies. For all its status as one of the developing world's great HIV prevention superstars, Thailand had until recently ignored the country's drug injectors. More than half of them were now infected with HIV. Just a year before the conference, Shinawatra had told police and local officials to be 'ruthless and severe' with drug suspects. There followed a wave of extrajudicial killings of junkies and dealers – over 2,500 deaths and disappearances were recorded in eighteen months.[35]

I lost my bet. Shinawatra did indeed mention junkies. He even mentioned that they had been neglected, and that the Thai government was 'now implementing a "Harm Reduction Program" to reduce the risk of HIV infection among injecting

* Some of the dresses were quite gorgeous. Check them out in the gallery section of www.wisdomofwhores.com.

drug users'. A group of Thai drug users in black T-shirts started heckling. They thought the government's policy amounted to murder, and they didn't think that quite counted as harm reduction. But they were on the upper terraces – Shinawatra probably couldn't even hear them. I, on the other hand, had control of what appeared on the 15-metre-high screens behind the diminutive Prime Minister, and I found it hard to disagree with the hecklers. My fingers were itching. With just a few keystrokes, I could have replaced the dancing elephant logo with 'He's a lying son of a bitch!' in letters two storeys tall. I didn't.

More daring was my colleague Karen Stanecki. She had been waiting to speak at the opening circus, after Shinawatra and Kofi Annan, after Miss Universe and the obligatory singing orphans. But by the time her turn came, there was almost no one left in the hall. During some interminable film in which young people behaved responsibly and hugged their friends with AIDS, the dignitaries filed out. Most of the other 13,000 people packed into the hall assumed the ceremony was over and followed suit.

Karen is an impeccably behaved UN bureaucrat with stellar diplomatic skills, but she had figured out exactly what was going on. It's de rigueur to have an infected person from the host country speak at the opening to an AIDS conference, and Thai activists had chosen Paisan Suwannawong, a young Thai injector who picked up HIV while he was in jail. The Thai organizers had a pretty good idea what he would say, and structured the whole show so that the hall would be empty and the live TV broadcast would have switched to the news before some HIV-infected junkie started saying ugly things about the government. Karen called their bluff. This well-mannered bureaucrat stormed up onto the stage. I mean *stormed*, silk shawl flying behind her. She grabbed the mike and invited Paisan to speak. Then she stormed off again. There were only a couple of hundred faithful left in the hall to hear Paisan. As soon as Thai TV figured out what was going on, they dropped the live broadcast from the conference and switched to Thai classical dancing.

Prime Minister Shinawatra, a former policeman, believed that you could effectively reduce the harm associated with drug-injection by cracking down on dealers and users. The populist politician has a keen sense of what plays well with the rural masses – he correctly judged that while middle-class urban Thais might find these summary executions distasteful, rural voters wouldn't punish him for what he sold as a 'law and order' approach to drugs. Ironically, the generals who removed Shinawatra from power in September 2006 are reverting to a more humane approach. Their government petitioned the Global Fund for AIDS, TB and Malaria for US$25 million, promising to spend a fair chunk on needle exchange programmes for drug injectors. Another example of a government taking a pragmatic public health approach to drug injection when it becomes less beholden to voters, who are so often squeamish about doing nice things for injectors.

Big pricks and small

Why are people so squeamish? There's the yuk factor of sticking needles into yourself for fun. Most of us don't care to think about that too hard. There's the stereotypical junkie thing, the image of grubby, long-haired youths skulking in dimly lit back-alleys littered with used syringes. And there's the chaos, the anxiety, the frustration, the disappointment that addicts trail around with them. It hangs like a cloud over their family and friends, their bosses and employees, it explodes into disruptive downpours at unpredictable times. I know a bit about this. While I was in Indonesia busily learning about addicts, my husband David was busy becoming one. He was taking party drugs, mostly, ecstasy and ketamine and coke, though pretty much anything would do. I suspect the morphine in the emergency medical kits he took to war zones didn't last too long.

David was a functional drug taker, just about. He still made it to work every day, and he's so good at his job that he could be pretty under par and somehow get by with a little help from his friends. The booze-soaked foreign correspondents of yore are a dying breed, but any hack over forty is pretty used to covering for colleagues who are 'tired and emotional'. Reuters had reaped the rewards of David's courage, hard work and exceptional touch with a human story from the battlegrounds of Iraq, Afghanistan, Ethiopia, Somalia, Congo and God knows where else for more than a decade, so the company cut him some slack, too. It was I who was most unforgiving.

I was painfully aware of the absurdity of the situation. I'll work day and night to collect good information from drug users. I'll spend my evenings on station platforms talking to addicts, trying to understand, trying to find ways to help. I'll argue until I'm blue in the face for programmes that keep junkies safe and healthy. But when addiction walks into my bedroom, I don't want to deal with it. I don't want to deal with the lying and the cheating and the parallel life in clubs with gorgeous, vapid girls half my age. The already infrequent visits cancelled for increasingly implausible reasons. The weekends that we did have together becoming detox clinics, passed between sweating, trembling and lots of sleep. I just don't want to deal with it.

With every non-weekend, David and I retreated further from one another. Finally in 2003, just after I'd been to Thailand to give a couple of presentations at the Fourteenth International Conference on the Reduction of Drug-Related Harm, we separated.

Happily, David has now dealt with his demons. Mine haunt me still. I know that it is often easier to be charitable to people who live in a different universe. That's why people walk past the beggar who sits at the door of the church and then transfer money to help victims of a tsunami in another continent. But still, I struggle with the contradiction in my own life. If I can't bring myself to be compassionate to an addict with whom I've shared ten mostly happy years of my own life, how can I hope to convince other people to care about a bunch of unknown junkie kids with tattoos and long hair?

I really can't. So we're back to the old tricks from Geneva days. Turn it into something people do care about. Throw up a smokescreen: if we don't deal with this it will become a development issue, become a security issue. Better yet, turn it into an 'innocent wives and babies' issue.

For a long time, many of us thought that was a non-starter. There's a widespread belief that junkies don't have sex. No sex means no onward transmission to people who don't inject drugs. Ergo, no problem.[36] On the one hand, I'd have liked this to be true. We had enough to worry about just with the injection. On the other hand, I secretly wanted to find evidence that injectors were having sex, so that we could make a case that preventing HIV among junkies would prevent a wider epidemic and protect the innocent wives and babies of the realm.

So it is that I am in my office at ten o'clock one evening in 2002, analysing data that have just come in from our injector survey in the port city of Surabaya. My colleagues have gone off to find drinks and listen to jazz. The data are a mess, not clearly coded, denominators not standardized. But I go straight for the jugular. What proportion of this group of 200 young, male injectors is having sex? Since it is Surabaya, home to one of South-east Asia's biggest red light districts, a city where even the tombstones act as nice flat surfaces for sexual transactions, I start with commercial sex. I'm half-hoping the percentage of injectors buying sex will be high; maybe as high as 10 per cent or so. I run the numbers: 20 per cent. That's bad news for the spread of HIV, though not so bad for my shameful secret agenda. The good news is that condom use is astoundingly high, close to 85 per cent. This is worth passing on to my colleagues over beer and jazz. I'm already out of my chair, just about to shut down my computer, when my eye falls on the scrappy code sheet. Wait, these data are raw, they are still coded 1 = yes, 2 = no.

When my data are nice and tidy, with the standard 0 = no, 1 = yes codes, my 'yes' row is always at the bottom of the table, so that's where my key percentage is. Because my data weren't coded, the results were 'upside down' for me, and I had read them wrong. Not 20 per cent buying sex but 80 per cent – the highest we have ever seen anywhere in the world. Not 85 per cent using condoms but 85 per cent having unprotected sex – right at the bottom of the global league table.* I slumped back down into my chair, my head in my hands. I admit I wanted some bad news. But I never conceived of anything this bad. Be careful what you wish for.

How did we ever get the idea that junkies don't have sex? That calm, sunny feeling that heroin gives you does eventually soothe the libido as well as the nerves. In men who have used heroin for a long time, testosterone and other sex hormones get squashed, and the sex drive goes with them. 'I definitely had less

* Besides buying sex, one in three of the Surabaya injectors had a wife or live-in partner, almost none of them an injector herself. Most injectors say their wives don't know they inject. One in four had at least one casual, non-paying girlfriend, and the majority of these guys had several girls on the go at once (Pisani et al., 2003a).

sex when I was on junk,' said Asoka, a languid Sri Lankan friend who has now given up heroin but is still an incurable dopehead. 'I did lots of dreaming. Though sex was very erotic when it did happen, as I remember.' Yogi Adrian agreed. But the longer he injected, the fewer opportunities there were for eroticism. 'Eventually, I wasn't even dating material because of the heroin. I started to think of heroin as my lover,' he said. He was quite surprised by how quickly his libido sprang back into action once he quit, though. 'Whoa. Where was I storing all of that?'

Most of what we know about heroin use and sex comes from studies of long-term injectors in industrialized countries. In a study in Texas, for example, just a fifth of longer-term users aged over thirty-five had sex with more than one person in the last year compared with over half of those who were younger, more recent injectors.[37] We don't really see this in the Indonesian data, but then most people haven't been injecting for long enough to tell.[38] The heroin revolution didn't come to Indonesia until the collapse of strongman Suharto's regime at the end of the 1990s, so injectors in the country's four largest cities had been injecting an average of only four years by 2005. More than a decade earlier in Amsterdam, the average time injecting was already fourteen years.[39] The libido of young Indonesian injectors has escaped relatively undented so far, perhaps because many also use amphetamines, which increase stamina.[40] I also suspect there may be an economic angle. Forced to choose between a drug high or an orgasm, most junkies will choose the drug high. But what if you don't have to choose? In Surabaya in the early 2000s, smack and sex were both dangerously cheap and many junkies were middle class. They could afford both types of high.

Not all of them, of course. In fact, some end up with the same double risk for HIV (injection and commercial sex) precisely because they are short on cash. They sell sex to buy drugs.

One of the reasons we were a bit clueless about the sex-and-injecting overlap early on in the Asian epidemic is that we simply didn't ask. When we were drawing up the surveillance cook-books in Geneva and Washington, we thought of 'risk groups'. Our 'respondents' were either sex workers, clients, drug injectors or members of some other group that fit neatly into an assigned box. We asked them about risks that we knew that particular group took – selling sex without a condom if they were prostitutes, buying it without a condom if they were clients, but not much else. This was in part because we wanted to keep questionnaires short. But it was also because we didn't think it all that important to ask about the 'peripheral' risks. So we didn't ask prostitutes whether they shot up drugs, or their clients if they ever had sex with other men. But as Fuad and so many others in Jakarta taught me, humans don't fit into discrete little boxes. It was precisely the people who indulged in more than one risky behaviour who carried HIV from one population to another. We started to change the questionnaires, not just in Indonesia but all around Asia. Soon, we were asking everyone about everything.

When we started to look, it didn't take long to explode the 'junkies don't get laid' myth.

In some parts of Indonesia, up to 4 per cent of female prostitutes say they inject heroin. Spread across the country, that would translate into around 8,000 women who both inject and sell sex.* Male prostitutes are even more likely to say they shoot

* The unholy alliance between commercial sex and drugs is a lot more pronounced in most industrialized countries than it is in Asia. At a conference in Northern Thailand, a UK-based colleague reported that crack cocaine was more or less a tool of the sex trade in Britain, making the sale of sex both more bearable and more necessary. She reported average earnings of £600 a week, about twelve times the amount a prostitute in Jakarta would earn. But £560 of that goes on drugs; we worked out that by the end of the working week, the two sets of women are in a similar financial position.

up. In Bangladesh, China, India, Pakistan and Vietnam, injectors are having sex. Lots of sex. And those are just the countries whose data sets are on my hard drive.[41] In the UK, much of Western Europe, Russia, North America, Australia, just about everywhere we have data, female injectors are far more likely to sell sex than women who don't take drugs. And selling sex is the easiest way to guarantee lots of partners.

Surveys in China are particularly telling. One in four of the men who inject drugs in China also shells out cash for sex. China is one of the few Asian countries where we have good data from female injectors. There, half of the girls who shoot up say they also sell sex. Injectors who sell sex are more likely to share needles than injectors who don't sell sex. That means the injectors who sell sex are more likely to get infected with HIV. But it doesn't stop there. Prostitutes who inject are especially unlikely to use condoms, and that means they can easily pass their infection on.[42]

So not only do injectors have plenty of sex, they have plenty of the types of sex most likely to pass HIV on to non-injectors. Don't get me wrong. I am not saying that a significant HIV epidemic among drug injectors will magically turn into a 'generalized' epidemic that will sweep through the heterosexual population and send other continents the way of Africa. Whether that happens still depends on who the injectors' partners have sex with, and whether they use condoms. What I *am* saying is that junkies plus sex will equal a bigger HIV epidemic than you would otherwise have had. And you'll see most of the difference in people who have never been near drugs.

Using new software designed to model the spread of HIV in Asian countries, I estimated that there would be fifty times as many people infected with HIV in Jakarta by the end of the decade than there would otherwise have been, just because we let HIV prevalence in injectors go from zero to 47 per cent in four years.[43] Two-thirds of people who we expect to die of AIDS because of our negligence are not injectors. They are a man who married a woman who used to be a sex worker who was infected by a client who had sex with a waria who shot up drugs. They are a woman who is engaged to a man who himself used to have a boyfriend who shot up drugs. They are a waria who had a client who bought sex from a woman who slept with a pimp who shot up drugs. In this little morality tale, clean needles for three people would have saved eleven lives, eight of them people who never touched drugs.

Even if you're sulky about drug addicts, you have to admit it would have been a good investment to give needles to those three people. Clean needles save lives, the lives of people's husbands and daughters, of airline pilots and doctors. The lives of people who may turn into your preacher, your lover, or your yoga teacher. Science, economics and compassion all dictate that we should help drug users stay healthy until they quit. Only our disapproval stands in the way.

8

Ants in the Sugar-Bowl

Ideology can sabotage sensible HIV prevention, no matter whether it is inspired by religion, politics or activism. It influences how money can be spent, and how it can't. It consigns poor old Fahdli to running sparsely attended peer support groups instead of giving injectors clean needles to cut HIV infection. It obliges Lenny to push a sex-free life to waria who pay their hospital bills by selling sex. It compels families of injectors in Thailand to collect corpses and organize funerals when they were hoping to collect groceries and organize a wedding. If you are trying to prevent an infectious disease, ideology is an all-round pain in the arse. But it is not the only thing that undermines sensible HIV programmes.

The other is money, plain and simple. The sheer volume of money now available washes away the need to use what we have well.

When I was sitting in Geneva banging the 'HIV-is-coming-to-get-you' tom-toms, poor countries were getting about US$300 million a year for AIDS, all told. We used to sit around fantasizing about having unlimited resources. Like winning the lottery. No more squabbling between researchers, between NGOs, between institutions to get their slice of the action. Then HIV started climbing the charts, both as a cause of death and as a *cause célèbre*. The number of people living with HIV rose by 38 per cent between 1996 and 2006. Spending on HIV in developing countries rose 2,900 per cent over the same period. By 2007 the world was spending US$10 billion a year on HIV in developing countries, the most that has ever been dedicated to a single disease.[1] Half of that came from governments of low- and middle-income countries (most of them in Latin America and Asia), as well as out of the pockets of the people affected by the epidemic. About 8 per cent comes from private foundations such as the Bill and Melinda Gates Foundation and the William J Clinton Foundation. The rest is given by taxpayers in rich countries.

The total continues to rise. The United States alone is budgeting US$4.2 billion for HIV in developing countries in 2008, more than double the US$1.97 billion it spent in 2006. A giant sugar-bowl of AIDS funding, and more sugar is pouring in every year.

As the Indonesian expression has it, where there is sugar, there will be ants.

The six founding members of UNAIDS have now swelled to ten, and ants from elsewhere in the UN system are making their way to the sugar-bowl, too. The Standing Committee on Nutrition has its HIV programme. So do peacekeepers. And fisheries. Why fisheries? Because, as a UN Food and Agriculture Organization research report explains: 'Fishing ministries must raise awareness of issues related to HIV/AIDS in the fisheries sector, initiate appropriate responses and coordinate responsive policies with health ministries.'* Chatting with my friend Bert who

* Food and Agriculture Organization, 2005. It is true that fishermen often have high HIV prevalence rates, because a high proportion of them buy sex. Whether the sex lives of fishermen is a matter for ministries of fisheries or the FAO is debatable.

works for the World Bank, I marvelled that almost every UN agency was on the AIDS bandwagon. Bert gave me a weary look. 'The UN institutions are professional beggars, and beggars go where the money is,' he said. 'So you get "culture and AIDS", "kids and AIDS", "fish and AIDS". I'm just waiting for "climate change and AIDS".'

This is no bad thing as long as we really have a problem with kids and AIDS or fish and AIDS. In Southern Africa, where we neglected the disease for so long that one in three adults has HIV, AIDS *is* a universal problem. You can make a plausible argument for needing to link AIDS to fish and culture in East Africa, too. But in most of the rest of the world there are only two issues, really: 'sex and AIDS' and 'drugs and AIDS'. If you don't want to deal with those things ('Not in my institutional mandate, guv') then you'd better butt out of HIV prevention.

Of course, that is not what's happening. All the rhetoric about HIV being a development problem has scattered so much sugar around that governments, institutions and NGOs can eat their fill without ever having to crawl into the bowls marked 'sex' or 'drugs'. The worst examples come from West Africa, where HIV is still concentrated firmly among those who buy and sell sex in most countries in the region, and where drug injection is surfacing as heroin trans-shipments through the region grow.[2] The HIV epidemics in West Africa may look a lot like some Asian epidemics, but the response is pure Southern Africa. Take Ghana, for example. Around 76 per cent of new infections still happen in commercial sex but 99 per cent of funding goes to 'general population' interventions such as micro-credit schemes and workplace outreach programmes.[3] In Nigeria, the Global Fund to Fight AIDS, Tuberculosis and Malaria has granted US$74 million so far for HIV, all of it for work with the 'general population'. Nigeria gets large slabs from PEPFAR, too, US$105 million in 2006 and rising; 90 per cent of the prevention money was going to 'general population' interventions.[4] 'Youth' is an especially popular focus for prevention efforts in Nigeria, even though HIV tests in several thousand recent graduates from technical college showed that just 1.2 per cent were infected – hardly a sign of an epidemic that is out of control among young people in the general population.

Meanwhile, Nigeria has a vibrant sex industry. I can't say how vibrant because the national programme has until now more or less ignored commercial sex. In a national survey in 2003, 3 per cent of men said they visited a prostitute in the last year, so that would be 1.2 million clients right there, and the probable total is a lot higher.[5] There are no estimates of how many women sell sex, and there's no routine HIV surveillance among sex workers. Sporadic studies are not encouraging. In 2003, 21 per cent of sex workers in the western city of Ibadan and 48 per cent in nearby Saki were infected with HIV.[6] Of course, we don't have a clue how much HIV is spread in sex between men in Nigeria, because no one has asked – the first studies got underway only in 2007. Scattered assessments in drug injectors in eight Nigerian cities in the early 2000s showed that they were as yet no more likely to be infected with HIV than non-injectors, which suggests that there's still a chance to prevent a major epidemic in this group.[7] But how much of the millions of dollars sloshing around for HIV prevention in Nigeria is being spent on drug injectors? As of mid-2007, none.

In Asia, it is slightly harder to get away with the 'HIV is everyone's problem' rap, because we've bothered to collect data showing that that is just not true. But the money available for AIDS in Asia is rising. The continent's share of rich-country funding leapt from an average of 11 per cent between 2001 and 2004 to 18 per cent in 2005 (largely at the expense of Latin America).[8] And we get substantial mismatches between where money is needed and where it is spent. In China, where 90 per cent of HIV transmission is in commercial or gay sex or drug injection, 54 per cent of all donor prevention money was paying for HIV prevention in the 'general population' at the end of 2005. In Thailand, where over 40 per cent of new infections were in commercial sex or among drug injectors in 2005 and where an epidemic among gay men was raging undisturbed, just 4 per cent of all HIV money went on preventing HIV during those very risky behaviours. Things were much the same in Cambodia – commercial sex and gay men accounted for 40 per cent of infections, but prevention in those groups was allocated just 20 per cent of the HIV budget.[9]

If you're asking for money in any other industry, potential investors would ask you to run the numbers. They want you to show how you'll maximize profits. In the case of HIV prevention, the profits are saved lives. You maximize them by providing effective prevention services to the people who are most likely to pass on an HIV infection, and those who are most likely to be exposed to someone else's infected body fluids. And yet none of the major funders asks us to run these numbers. The biggest funder of all, the United States, decides in Washington how its money will be spent. It actively discourages investing in ways that will maximize lives saved in most of the world. Others such as the Global Fund invite countries to make their own decisions. Countries have to show that there are HIV positive people on the project design team. They have to show that government is in partnership with 'civil society'. They have to show that they are being inclusive of 'vulnerable groups'. But they never have to show that they will prevent the maximum number of infections for the minimum number of dollars. The result is a colossal waste of taxpayers' money.

Fashion victims

I learned just how colossal when I started hopping from Indonesia to neighbouring East Timor. This minuscule country is perched on the eastern half of a rocky island north of Australia; the other half belongs to neighbourhood bully Indonesia, which invaded the east in 1975, beginning a low-grade guerrilla war that ran for more than two decades. A referendum on independence in 1999 sparked a conflagration between pro- and anti-independence groups, and the Indonesian military enthusiastically threw petrol on the fire. Most of the territory's infrastructure was destroyed. UN peacekeepers intervened and East Timor eventually rose from the ashes in 2002, the first independent nation to be born this century. Christening presents were in order. Australia was generous – it has given over US$400 million since independence, a lot of it in long-term projects that aim to build the nation's strength bit by bit. After fratricidal conjoined sibling Indonesia, Australia is East Timor's nearest neighbour; the two countries have a fragile agreement about potentially lucrative natural gas fields under the sea that divides them. East Timor often seems to totter on the brink of chaos. If it falls, Australia will get splashed. So it is understandable that Canberra wants to help its neighbour grow up strong. The United States, on the other hand, has no strategic interest in this barren little scrap, tacked on to the outer petticoats of Indonesia. It seemed polite to give the new nation a christening present, certainly. But just the once. Washington did not want to get locked into stumping up a birthday present every year.

So America gave East Timor the season's most fashionable development item: an HIV programme.

The State Department put a million dollars on the table for HIV prevention in East Timor, and USAID later topped it up with another million. At the time, Africa was spending around 51 cents per adult on HIV prevention and care, or US$15 per HIV-infected person. This gift would give East Timor thirty times as much per person for prevention, and a whopping US$285,700 per known HIV infection. Because when the US came up with its two-million-dollar present, only seven people had ever tested positive for HIV in East Timor. Seven.

To those of us working in the region, this seemed totally absurd. East Timor had the highest birth rates in South-east Asia, and the lowest immunization rates. Mothers and infants both died more frequently than anywhere else in the region. But there was no indication that HIV was a threat. We had never heard any reports of drug injection, and when the territory was part of Indonesia, even the prostitutes had been imported into the fiercely Catholic little enclave to serve civil servants from other islands. When the Indonesians left, the few red light districts went with them.

But Timor came of age at the height of the 'HIV is a development problem' boom. The new government had been sold a line, and it went like this:

1) HIV is a development problem. East Timor has lots of development problems. Ergo, East Timor will have lots of HIV.
2) Cambodia is in Asia. It was torn apart by a nasty conflict, and then the UN peacekeepers arrived. It now has a big AIDS epidemic. East Timor is in Asia. It was torn apart by a nasty conflict, and then the UN peacekeepers arrived. Ergo, East Timor will have a big AIDS epidemic.[10]

The eminently sensible health minister, Rui Maria de Araujo, could not resist the onslaught. In a statement released on his behalf by the United Nations Development Programme, he wrung his hands about the danger of an HIV epidemic: 'All the necessary ingredients are there – poverty, street children, joblessness and prostitution.'[11] The country named HIV as one of its top four health priorities. When the US suggested its US$2 million HIV gift, it was welcomed.

Does that mean East Timor got US$2 million to spend on HIV prevention? Of course not. America doesn't actually give money to foreign governments. Instead, it puts it into the development food chain.

At the top of the development food chain is the United States government, which gives away US$23 billion of its taxpayers' money every year for all manner of development projects.[12] Some of that goes to the United Nations or the Global Fund and they give it to governments. But most gets earmarked for projects decided in Washington, in consultation with US embassies around the world. Each proposed project get posted publicly, and anyone who can figure out the positively byzantine US government tender systems can bid to take the money and run the project.

Most of the bidders are Beltway Bandits – organizations which squat around the Washington DC ring-road, never far from the comforting nipple of USAID. When the government announces a new contract, the Bandits leap into activity, writing proposals that show how cleverly they will spend the money. Or rather what's left of the money. The Beltway Bandits slice between 15 and 30 per cent off the top before they ever pass it on to a country office, presumably to pay for the money spent bidding for proposals that keep people like me in a job.

If Washington's original project plan makes sense, the process isn't too painful. You write a proposal that you think will make a difference and hope yours wins. If it does, you take your 30 per cent cut and then get to work making a difference.

But what if the original project plan makes no sense at all? What if you know that what they're asking for will make no difference? Then you've got a problem. Here are some options:

a) Don't bid. No 30 per cent, but no time wasted doing daft things that make no difference.
b) Do what the donor wants. Win the bid. Take your 30 per cent. Spend the other 70 per cent making no difference.
c) Bait and switch. Tell them you'll do what they want. Win the bid. Take your 30 per cent. Throw your agreed plan away and try to do something that will make a difference. (This is a headache. Keep lots of aspirin handy.)
d) Be stubborn. Tell the donor that their plan won't make a difference, and make a compelling case for a plan that will. Probably lose out to someone who has chosen option b). Just possibly, win. Do something that makes a difference. (This is your Best of All Possible Worlds. Keeps lots of champagne handy.)

The 'stubborn' option is based on the premise that most people in the donor agencies are also smart, dedicated people who want to make a difference. If their plan doesn't make that much sense to those of us who are working in brothels and bars, in city hospitals and district health departments, it is often because they are further from the facts than we are, and closer to the dominant political ideology. If we make a good case that changing the plan a bit will save more lives for less money, then the funding agency will usually go along with it. But not always.

When the US government announced its US$2 million HIV gift to East Timor, the Beltway Bandit that I worked for, Family Health International, didn't even have a crack at option a). The

Timor project was essentially funnelled into an existing contract between FHI and USAID. We just had to deal with it.

The Asia staff started a bidding process of our own. I suggested that we should round up every prostitute in the country and send them all to Harvard. We'd have enough money. That idea was seconded by Jeanine Bardon, a blonde bombshell of a virologist who was head of the Asia programme. She has managed to retain both a passionate commitment to doing what works and a sense of humour – not an easy combination when you are running a US$76 million programme across twelve countries. If Philippe (the Gérard Depardieu look-alike who runs the gay boy programmes) had been bidding, he'd have probably suggested sending them to university in France – so much cheaper that we'd be able to send every gay party-boy in the country, too. Nancy Jamieson, whose speciality is behaviour-change communication, suggested skipping all the communication nonsense and just dipping every potentially sexually active person in Dili in latex. 'I am not sure how decisions are made in Washington,' she wrote in an e-mail. 'Spirit guidance? Dowsing rods? Lucky-pick?'

In the end, though, we had to come up with some kind of sensible proposal to spend US$2 million on a problem that we all suspected didn't exist. I went to the Timorese capital Dili in 2002, looking for trouble. Diligently, I chatted to cabbies, cops, cigarette vendors – the kind of people who usually know where the action is. I trawled the cafés and bars. It was lonely work.

Sitting at the bar at Poy Chollor, the biggest nightclub in East Timor, I finally found a companion. It was around midnight on a Thursday, and the one other person at the bar was a soldier from Mozambique. He stared at his warming beer, disconsolate. 'I thought this place would be like Cambodia,' he mumbled. I thought back to the early 1990s, when white Toyota pick-ups were parked three deep in the rutted roads of Phnom Penh, having disgorged their Blue Berets into any one of the capital's teeming bars. Soldiers sat pink under the neon, a golden girl hanging like a ripe peach off each arm. Around them flitted more girls, pressing forward cold beer. If Dili was being sold to peacekeepers as a second Phnom Penh, someone should complain to the Advertising Standards Authority. Dili's handful of bars tottered into life only at weekends. Outside the bars there was no action at all. 'Forget about it,' snorted my Mozambican bar-mate. 'Here, you touch a girl and whack!' he hacked the blade of his flattened hand into his crotch. He wasn't tempted by the smattering of local prostitutes who slouch along the seafront promenade, and it's not hard to see why. Toothlessness is not a prime draw in a lady of pleasure.

'All the ingredients for an HIV epidemic', the health minister had said. And the ministry had busily printed thousands of posters to keep this epidemic at bay. Bright green, with a red ribbon and the illuminating slogan: 'Everyone is at risk.' The poster appeared on the door of the HIV testing centre set up by a Dutch NGO right next to the chronically overcrowded polyclinic in the city's main hospital. There was another poster, too, of school kids and housewives, farmers and sailors all trooping off to get an HIV test. But none of the noisy crowd waiting for the polyclinic to attend to their putrid wounds and cataracts and intestinal worms wanted an HIV test. The service had only had three clients in its first six months. Meanwhile WHO was teaching workers at primary health care facilities around the country how to deal with sexually transmitted infections. It would help keep the lid on the explosive HIV epidemic that must be on its way.

This all seemed a bit premature to me. Poverty, street children and joblessness do not make an HIV epidemic. From what I could see, there wasn't enough prostitution in East Timor to do the job, either. No evidence of any drug injection. Among the seven people with diagnosed HIV infection, three were in one family (man, wife and infant) and a couple more had contracted their infection while living in Indonesia. Nearly 1,500 surveillance tests, including in the East Timorese military, had found zero infections. For a while we toyed with the idea of doing a laboratory study that would lead to national treatment guidelines for sexually transmitted infections, but the clinics and hospitals saw too few STIs to make the study possible, let alone worthwhile.

In the end, we proposed a rearguard 'bait and switch' approach. We'd do a really solid study to measure HIV, other sexually transmitted infections and risk behaviour. And if indeed we found none, well, we could at least hope to stop funders throwing good money after bad. There was no shortage of things East Timor really did need, after all.

We tracked down most of the sex workers in Dili (about 100 in all), as well as a fair few of their clients, whom we found among taxi drivers and young soldiers. We talked to most of the gay guys who cruise the parks and the Saturday night discos. We stuck needles into people and asked them to pee into jars. We took swabs from all manner of orifices. We asked people questions about their most intimate behaviours. And we found virtually nothing. Four of the prostitutes – three women and one transgender – were HIV infected; they were among the few who reported sex with foreigners. None of the clients we tested was infected, which suggests that the virus was not doing the rounds in indigenous commercial sex networks. None of the gay guys was infected. The prostitutes reported so few clients that we had to change the questionnaire to ask about clients in the last month rather than the last week. Condom use was pitifully low, but it was still impossible to puff our findings into any kind of impending storm clouds.[13]

The study itself was actually a pretty good use of money. It proved definitively that *not* everyone was at risk. It was also a relatively quick way to make contact with most of the people who had even a remote possibility of exposure to HIV in Dili, and to set up basic services for them. (We found that not a single one of the sex workers we interviewed had ever had an STI check-up before the survey, and 40 per cent of them had never seen a condom before.) The study identified a need for easier access to condoms and lube for gay guys, too. But if East Timor did just those few things in the capital city and along the border with Indonesia – condoms, lube and STI screening and treatment for a few hundred people with well-defined risks – it would pretty much lock out the possibility of an HIV epidemic completely. And it wouldn't cost anything like two million dollars.

Well, what was left of two million. Head office had taken its tithe, and we had to kit out an office (improbably shared with World Vision, who sang hymns each Friday morning as we ran around on the other side of the nose-high partition screeching about vaginal swabs and lube – it quite took me back to my Geneva days). There were staff to hire, and technical consultants like me to pay for, and fees to pay to get our medicines out of customs – it put a fair hole in the available cash, and *still* we had more money than was needed to shut down the threat of HIV in East Timor. Now, at least, we had the data to prove it.

We went to see the US ambassador in East Timor, Grover Joseph Rees. Rees was one of the architects of the famous President's Emergency Plan For AIDS Relief; it may be no coincidence that his appointment came along with a slab of money for AIDS. A former henchman of Chris Smith, the Congressman who brought us the anti-prostitution loyalty oath, Rees knows how things work on Capitol Hill. We asked if the funds could be redirected, to immunization, for example? 'These things are decided by a bunch of guys sitting around on committees in Washington. There's not much we can do about it from here,' he said. I persisted. 'Wouldn't it make more sense to spend money on diseases that exist, rather than ones that don't?' I said. I laboured the point: 'It's good economics, as well as good public health practice.' The ambassador drew his hand heavily over his face. 'Every profession is the same. They think they have the knowledge and everyone else just has ideology.' He'd clocked me, all right. But still, I thought it was a bit harsh when he went on to paraphrase the Nazi propaganda minister Paul Joseph Goebbels: 'When I hear phrases like good public health practice, I reach for my revolver.'*

Rees had different ideas about how to spend what was left of the money. He hadn't chatted to any Timorese prostitutes, but he *had* been talking to a fellow American called Dan Murphy, who ran a clinic in a desperately run-down area of Dili known as Bario Pite. Doctor Dan's clinic drew queues of people every day. Pregnant women and ailing children, men on crutches and teenagers with gunshot wounds, they all passed through the clinic as if on a conveyor belt, paying what they could, not at all if they couldn't. The clinic was subsidized by Christian groups in the United States, and in Ambassador Rees, Doctor Dan had found a natural ally. He'd had one patient with AIDS, and suspected that another might need treatment. And he was lobbying for drugs. Rees had a question: with all that money out there for

* Meeting on 26 September 2003. At the end of the meeting Rees said, 'I have not said anything in this room today that I would not be happy to see on the front page of the *New York Times*.'

HIV, why couldn't we just give Doctor Dan a stock of antiretrovirals?

There were so many reasons why not – medical, ethical, legal, financial, logistical – that it took a five-page briefing paper to explain them all.[14] The reasons were not good enough for the ambassador. 'We have all this money for HIV and our country makes all these drugs. We can't just let these people die,' he said. I could see his point. On the other hand, we could save thousands of lives if we spent the same money on far more common diseases. 'I know that, but I believe I will be judged on how I deal with the people who fall into my path,' Rees said. He had summed up one of the great difficulties of public health prevention: helping unseen masses stay disease-free is less tangible than giving medicine to people who are already suffering. Prevention doesn't deliver the same feel-good sense of compassion. And it provides fewer photo-ops.

The biggest obstacle to helping Doctor Dan's one-maybe-two patients was actually buying the damned drugs. Procurement is an enduring nightmare for people working with US funds. That's because that great, free-trading, open market economy of the United States has laws that channel its overseas development dollars back to companies at home. If you want to give out wheat to starving people or infant formula milk to the babies of HIV-infected women, you have to buy them in the US. Condoms, computer equipment, cars – they all have to come from US manufacturers at US prices. USAID has bought some nine billion condoms for family planning and HIV prevention programmes since the mid-1980s. Most of them come from Alabama, at a cost of around five cents each.[15] The American taxpayer then pays to ship them to countries like Indonesia, India and China, where they would have cost two cents a piece to buy. Americans have spent US$270 million more than they needed to on condoms – that would pay for a lot of retraining for factory workers in Alabama. And it doesn't include shipment costs, which are substantial. The 'Buy American' policy has produced some wonderful carbon footprints. A condom used in East Timor and made in Alabama from Sumatran latex has travelled 20,200 miles.

With drugs it is even worse, because they have to be approved by the US Food and Drug Administration. That puts lots of drugs that are cheap, practical and widely used in developing countries off limits. Instead we have to buy sophisticated US formulations which require cold storage. In the East Timor study, for example, we were testing for syphilis, and offering free treatment for any cases we found. The standard treatment, bezathine penicillin, costs about US$1.50 a dose in neighbouring Indonesia. And it's not particularly sensitive to temperature, so I could just buy a couple of boxes and stuff them in my hand luggage. I didn't expect to find all that much syphilis, so US$100 should do the trick.

But no. Because of the 'Buy American Act', which dates from 1933, I have to shell out US$71.15 a dose for some product in fancy-schmantzy packaging, made in Tennessee. And then I have to get it from Tennessee to a clinic in the sweltering dust-bowl that is Dili, keeping it between four and eight degrees centigrade all the while. My hundred bucks is starting to look more like 10,000.[16]

If the idea of paying 100 times more than you need to for medicine gives you a stomach ache, there is another solution: horse-trading. A lot of countries don't funnel their overseas aid money to the captains of industry back home. The UK outlawed such favouritism in 2002, after its overseas development programme was hived off from the Foreign Office; Ireland doesn't tie its aid either. Most Northern European countries tied less than 5 per cent of their aid in 2005.* So you do a deal with a programme funded by the Brits, or some other donor who doesn't care where things come from as long as they're cheap and do the job. You pay for my syphilis treatment, I pay for your training programme. We do the same kinds of deals to get around ideological restrictions, too. In Indonesia, Britain and Australia are buying needles for drug injectors who are being reached by US-funded programmes. Organizing these deals sucks up time and energy and makes you feel a bit like a used car salesman. It also skates along the edge of legality. But it certainly saves a lot of tax dollars, and that makes for a lot of blind eyes among the people in USAID's overseas offices, who tend to care more about lives in the country they live in than jobs in Alabama.

The one thing it is really hard to get hold of in an under-the-table deal is antiretroviral drugs, the very thing the US ambassador in East Timor wanted. When Bush first announced the President's Emergency Plan For AIDS Relief, he earmarked US$10.5 billion for HIV treatment and palliative care. All the drugs would come from the US, naturally. We joked that PEPFAR actually stood for 'Purchasing Expensive Pharmaceuticals From American Retailers'. As if to confirm our cynicism, Bush appointed the former head of pharmaceutical giant Eli Lilly to head the initiative. A lot of the first generation antiretrovirals were closing in on the end of their patents when PEPFAR was announced. Drugs that cost US$10,439 for a year's supply as recently as 2000 had crashed in price. This was in part

* The worst offenders were Canada (33 per cent), Australia (28 per cent) and the countries of southern Europe, especially Greece (26 per cent) and Portugal (25 per cent). The US has not reported the status of its aid to OECD since 1996, but when it last did, 72 per cent of its aid was tied. That excluded technical assistance, which is almost all provided by US institutions. Source: OECD Development Assistance Committee database.

because brave Brazil stuck two fingers up at various rump patents, making generic versions of the drugs at US$ 2,767 per person per year. Then India waded in at just US$350 a year. These prices weighed heavily on manufacturers in the West, who were already down in the mouth because decent HIV prevention was eating into the potential market for their drugs in rich countries. The number of new AIDS cases was falling dramatically. The astronomical 'need-to-recoup-our-investment-in-research' prices stumbled and fell too. By the time PEPFAR started, a combination of three US-made drugs cost just US$727 a year – only four times as much as the cheapest generic equivalent.[17] Big Pharma was badly wounded in the AIDS treatment battle. Then along came this gift from Congress – PEPFAR would guarantee an extra two million patients for US drug manufacturers.

'We could treat four times as many people for the same amount of money if we were allowed to buy generics,' treatment activists yelled. And yelled, and yelled, while Washington talked about quality, and safety, and maintaining standards and . . . and . . . and . . . until they finally ran out of excuses. There are still all manner of restrictions on what drugs you can buy with PEPFAR money, but pressure groups have scored monumental victories. By March 2007, 88 per cent of antiretrovirals bought with PEPFAR money were generic.[18]

One of the reasons why it was difficult for us to buy drugs for Doctor Dan's patients was that the American's one-time christening gift to East Timor was scheduled to wind up the following year. HIV drugs are not just for christenings, they're for life. You can't just start someone on the drugs and walk off.

The more donors focus on providing treatment, the more worried we should be about abandoning patients. There were about two million people on antiretroviral treatment in the developing world by the end of 2006, getting on for a third of those who needed the drugs.[19] That's pretty impressive considering that no one was even talking seriously about treatment in the poorest countries when I was working in Geneva less than a decade earlier. But it presents us with a new dilemma. Right now, AIDS is in development fashion and we're flush with money for treatment. But what happens when the Global Fund abruptly turns off the tap on a country because it suspects corruption in one part of one project? What happens when PEPFAR comes to an end (it's an 'emergency response' after all)? What happens when other big funders waltz off to try on the next fashion item – global warming, almost certainly? I don't know how soon this will happen, but it will happen sooner or later. Who will go on buying medicine for the millions of men and women who need it? Few governments seem to have thought that far ahead.

Spoiling the broth

East Timor is a tiny fish in a vast sea of HIV funding. Plankton, really. But its story is repeated on a grand scale elsewhere. It was in East Timor that I had to abandon a mantra that I had been chanting with increasing frequency over the years: 'Relax! There's enough HIV for everybody.' It first popped out of my mouth in Kenya in 1998, as I watched NGOs and researchers squabbling over HIV projects. Of course what they were squabbling over was not HIV but money – at the time, there wasn't enough of it. But as the sugar has poured into the AIDS funding bowl, everything has been turned inside out. More and more organizations have swarmed into the bowl, competing for funds, competing for staff, competing for projects, competing for the time and attention of governments. The US christening present wasn't the only one to divert the attention of East Timor's infant government from more important things. The Australians had an HIV project, and then there was UNICEF, the World Health Organization, Catholic Relief Services and a bunch of other NGOs too. If you scratched around enough, you could probably find more organizations working on HIV in East Timor than there were people infected with the virus. Suddenly, there really wasn't enough HIV for everyone.

Even in much bigger countries with much bigger epidemics, the market traders of the HIV industry are shouting one another down to get contracts.

Not a bad thing, you might think. A bit of good, free-market competition is just what these bloated NGOs need to shake them up and increase efficiency. And you'd be right. Except that the AIDS industry isn't a free market. As we've seen, you rarely have to say what your 'bottom line' is – how many infections you'll prevent. And you almost never have to show you've prevented any infections. You can be judged a success for just doing what you said you were going to do, like build a clinic, or train some nurses or give leaflets to 400 out of the nation's 160,000 drug injectors. It's a bit like declaring that Ford is doing really well in the car market because they've got factories and floor managers and an advertising campaign, instead of looking at sales figures. Or even checking that they make cars that run.

In the AIDS industry, most people do whatever projects pay them to do. 'In fact, we don't have time to do any real work,' protested a Chinese colleague, whom I'll call Wang. He was from the local health department in Dali, a Wild West town on the opium-sodden border between Yunnan and Burma. Huge casinos just inside Burma acted as a magnet for gamblers from all over China, and gamblers acted as a magnet for prostitutes. Wang had been pulled out of Dali to be trained in how to estimate the number of prostitutes locally, and he was quite cross about it. He'd already counted prostitutes in Dali three times in the previous year, he said. Once for the China–UK prevention programme, once for a US-funded programme and once for the Chinese government. Now he was being asked to do it again with Global Fund money. 'Count, count, count. And no money for prevention.' Wang was getting louder as he got more worked up. I could see why he was upset. 'Why don't you just give all the donors the same count and have done with it?' I asked. Wang looked shocked. 'But they've all given me money to count!' he bellowed. 'If I didn't count for each of them, that would be corruption!'

Wang was a victim of one of the great hypocrisies of the AIDS world: something called the Three Ones. In public, all the biggest donors wring their hands about the need to coordinate. On paper, they have all agreed to work together in each country, dealing with only One national coordinating body, One national plan for HIV, One monitoring and evaluation system. In other words, Wang, Yunnan province and even all of China only needs to count anything once, through its One national surveillance system (which is the core of the One monitoring and evaluation system).

Everyone prays to this Holy Trinity in public, but most honour different gods in private. With a bare handful of exceptions, most notably the UK's overseas aid outfit the Department for International Development (DfID) and some of the Nordics, everyone wants to be able to account for what was done with *their* one dollar's worth of programming. The Global Fund wants to be able to say, as it does on its homepage as I write: 'Global Fund Helps Deliver Sharp Increases – Over 1 Million on AIDS treatment'; George Bush wants to be able to say, as he did in his speech announcing a second wave of PEPFAR: 'The Emergency Plan has supported treatment for 1.1 million people infected with HIV.'[20] It is a massive waste of time and energy to measure and report the same things over and over again, let alone to try and disentangle whose condom contributed to which prevented infection. And yet you can see why it happens. I am the first to rant about lack of accountability when people can't measure what they have delivered. Donors are damned if they measure and damned if they don't.

Too many AIDS programmes spoil the broth for the people like Wang who are just trying to get on with delivering services to people who need them. And the problem is getting worse all the time. PEPFAR, for example, works mostly through Beltway Bandits and church groups. It won't give more than 8 per cent of its money in any one country to a single organization. That's a minimum of thirteen separate first-line contractors getting US money in any given country. That's thirteen country directors, with nice rented villas and kids in expensive international schools. It's thirteen offices, thirteen finance directors, thirteen janitors, thirteen photocopiers. It's thirteen separate organizations trying to hire the best and the brightest away from the national institutions they work in, and thirteen organizations trying to set up capacity building workshops to replace the skills they've stolen. Each organization is clamouring for the time and attention of whoever's left in the national AIDS programme, and each is trying to pass money on down the line to the few local institutions and NGOs who can actually do what's needed on the ground. The good NGOs might get money from three or four of the Bandits, and they'll have to write separate monthly, quarterly and annual reports for each of them. In the best of all possible worlds, the naked inter-agency competition is tempered by a sense of a common goal. As one weary-sounding colleague describes the result, he reminds me forcefully of Wang: 'We can't get any work done. We're all too busy coordinating with one another.'

That's the best-case scenario. Even worse is when huge chunks of money go to organizations that *don't* open offices in the country where they are supposed to be working. I've been pretty rude about the Beltway Bandits, including my own employers. But in fact, I was proud to work for FHI in Asia. Most of the staff are Asian, and while a lot of the top jobs and regional technical positions go to expatriates like me, almost all are people who have chosen to live in the region for huge chunks of their adult lives, who work in the local languages, who have some clue about the cultural niceties, and who make time to learn from friends and colleagues who understand the situation better than they do. FHI has been on the ground in Asia for over twenty years. Its staff know how to get things done.

They've even become expert matchmakers. Under the guidance of the blonde bombshell Jeanine, the organization learned to marry money that came with ideological strings attached to money that was more promiscuous. We bought sterile needles with money from the Global fund or DfID, and married them to outreach distribution systems paid for by USAID. The offspring: programmes that do what is needed on a scale that might even begin to make a difference. Several other Beltway Bandits share these skills. Condom promoters Population Services International and research and policy nerds The Futures Group are among those I've admired in the course of my work. Yes, we all do our share of running around the sugar-bowl, but at least we're delivering services to people on the ground. And we've been around long enough to make a lot of mistakes and learn from them. When we looked at the pathetic number of people we were reaching with our Indonesian programme in 2003, we made huge changes, reducing the emphasis on one-to-one communication and introducing incentives for local NGOs to expand their reach. When we realized that you can't do effective

HIV prevention for injectors without offering needles, we matchmade our way to a solution.

But the piles of new sugar have brought lots of new ants. Many of them have no experience outside the United States, and few have much experience working with hookers, rent boys or injectors. There's a lot of very expensive 'technical assistance' – people flying in from universities in distant New England for a few days to attend workshops and meetings in languages they don't speak. They dispense their pearls of wisdom and fly off again, leaving people in the country to figure out what to do about it. The bigger the programme the more ants there are. In Nigeria there are thirty-seven different organizations sucking in cash just from American taxpayers, and a whole lot more feasting on funds from other sources. Those first-line pan-handlers pass the money on to hundreds of local NGOs to do the work; inevitably they end up fighting over who gets to work with the best organizations.

The NGOs can get pretty competitive too. In the dingy outpost of Merauke, in Indonesian Papua, I was visiting the office of an NGO that provides care to people with HIV (PWHA, or People Living With HIV/AIDS, in the clumsy industry jargon). One of the care workers came in in tears. I tried to find out what was wrong, but she wasn't very coherent. 'She was my PWHA, mine!' she wailed. When she calmed down I got the whole story. She was upset because a 'rival' NGO had 'stolen' one of her HIV positive clients.

The push factor

If NGOs are squabbling like jealous children in a family that has grown too big, why don't the big donors just knock their heads together?

Because they are in competition too. Each one wants to curry favour with the governments they are giving money to, each one wants to be seen as the good guy so that next time a big oil exploration or cell phone contract comes up maybe their businessmen will have an edge. Britain was a champion of the aid–trade two-step from the time Margaret Thatcher came to power in 1979 until 1997, when the UK's overseas aid administration was liberated from the Foreign Office, and the Department for International Development was created.

Now Britain has changed its tune. Together with some of the Nordic nations, it is trying to lead the world away from the do-as-I-say-not-as-you-need bilateralism lately more typical of the United States. It gives money to governments that it believes have developed sensible programmes, and lets them get on with distributing it as efficiently as they can. London also channels a lot of its money into so-called 'multilateral' programmes, such as the Global Fund, WHO and UNAIDS, that many countries support. Theoretically, if everyone gave their money to these organizations instead of setting up separate programmes they can wrap their national flag around, the diligent Wang in Yunnan would have to do a lot less counting.

But giving up control of how aid money is ultimately spent has its pitfalls, too. It is hard to monitor how governments on the receiving end actually spend the cash. And it makes it very difficult to answer to the chorus of NGOs, voters and lobbyists at home. The lyrics in London might go something like this:

BBC INTERVIEWER: 'So you're saying we don't have an AIDS treatment programme in South Africa? I thought there were more people with AIDS in South Africa than anywhere else in the world.'

DfID OFFICIAL: 'Well yes, but South Africa's quite rich – they really don't need that much help. In any case, our American friends have given them half a billion dollars in the last three years, and half of that's for treatment. And we are among the largest contributors to the Global Fund – they're supporting South Africa with another 122 million. At DfID, we feel that British tax money would be better spent . . .'

OXFAM REPRESENTATIVE: 'It's outrageous! AIDS is *the* major development issue of our time. We have the money and the medicine, they have the need. How can we not provide treatment? It's just unethical.'

DfID OFFICIAL: 'It is DfID's policy to work with governments and other partners to avoid duplication, so that we can achieve the greatest reduction of poverty in the most cost-effective way.'

JOE BLOGGS (listening to the radio over his cornflakes): 'Plonker! Those Westminster types, they bang on and on about Africa, but do they want to buy medicine for people with AIDS? Not a chance. Bloody typical. We need more people like that Bono, that's what we need.'

Actually, what we really need might be more people like Bill and Melinda Gates and squillionaire Warren Buffett. These New Philanthropists have the potential to change the face of international public health, because they have gobs of cash and no voters to answer to. Of course, they still have to play nicely with governments in the countries where they want to work, but they are in a better position than most to tackle projects that governments are nervous about taking on themselves, and to do them on a scale that might make a difference. The Gates

Foundation's US$258 million Avahan project in India focuses largely on sex workers and their clients, as well as drug injectors. They are giving out eight million condoms a month and providing STI screening and other services for groups that have been neglected by the government but that account for most of the new HIV infections in the country. In the two northern provinces with the highest rates of drug injection, the Foundation and its partners are handing out around 271,000 needles a month to more than 16,000 injectors.[21] That's more than a Prada boutique-sized effort. It is early days, but it looks like these initiatives might actually make a difference.

The Global Fund to Fight AIDS, Tuberculosis and Malaria was also designed to sidestep the headaches of bilateral aid. It was conceived as an assault on the mushrooming of programmes driven by donors and their various ideologies, programmes that duplicated one another, left huge gaps, and obliged governments in poor countries to spend more time measuring and reporting things than actually doing them. The idea was that everyone would put their money into one big, happy pot. Countries would form consortia that had to include people from inside and outside government, as well as people with HIV who wanted to take part if they could find any (East Timor couldn't). These groups would decide what the country needed, and make a bid to the Fund. The bids would be reviewed by scientists not ideologues. The proposals most likely to make a difference to the national epidemic would get money. The rest would get sent off empty handed.

UN Secretary General Kofi Annan first called for a 'war chest' to fund a prolonged battle against AIDS in early 2001, during his 'Wall of Silence' speech to African leaders – the one that didn't mention sex. A month later, the United States, Britain and France between them pledged half a billion dollars to kick off a fund. Within a year, the Fund proper was born into a seething cauldron of expectations. It professed to follow Three Commandments: 'Raise it, Spend it, Prove it'. By that time, US$1.5 billion had been raised and the pressure to spend it was huge. Countries were given a paltry five weeks to put together their consortia, agree on what was needed and get their proposals out. Close to 400 bids were thrown into the ring in that first round, covering three very different diseases. The whole lot was reviewed by just seventeen people in three sleep-deprived days. In the end, only thirty-six proposals were left standing. Indonesia's HIV proposal was one of the victors.

Indonesia had said it would provide services for drug injectors and sex workers in four provinces, as well as running a bunch of mass media spots and increasing knowledge among 'youth'. The proposal included all the must-have phrases: 'scaled-up', 'best practice', 'multisectoral response', 'empowerment' – a full card for Bullshit Bingo. (Bullshit Bingo is a game we play in really, really boring meetings. You make a card with all the push-button development-speak phrases and then check them off every time a speaker trots one out. The first person with a full card wins. But to win, you have to stay awake.) Yes, the Indonesian proposal had all the push-button words. But cut to the chase, the anticipated results that might actually lead to less exposure to HIV: peer outreach for just 400 of the nation's 160,000 drug injectors, antiretroviral treatment for eighty people, and some condoms for prostitutes.[22]

Had the Global Fund's technical review panel had more time and more sleep, they might not have invested US$7.8 million in a proposal with such flimsy goals. Or perhaps they would. In the next round of bidding, when things were a bit calmer, the technical reviewers saw fit to grant Bangladesh nearly

US$20 million for a programme among young people to 'help avert a generalized HIV epidemic in Bangladesh'. The likelihood of a generalized epidemic among young people in Bangladesh is zero. I can't give you any stats on the distribution of HIV in Bangladesh because at the time of the grant the country's very efficient surveillance system, which covers all the highest-risk groups, had found no more than a handful of infections among drug injectors. There was some risk – besides the drug injection there is an active commercial sex scene and a fair bit of anal sex between men. But this risk wasn't ever going to be prevented by education campaigns among young people in the general population, and the Global Fund must have known it.

It didn't seem to matter. The Fund needed to push money out the door, and fast. The Second Commandment (after 'Raise it!') was 'Spend it!', not 'Spend it well'.

We fantasized for a while that the Third Commandment, 'Prove it!', might mean 'Prove that you've spent it well.' But no. It just meant 'Prove that you've spent it according to the rules.' The rules allow you to claim that you'll dent the Indonesian HIV epidemic by chatting with 400 drug injectors and reducing the viral load in eighty people. Everyone knows that's just silly, but the Global Fund has given you US$7.8 million for it anyway (that's why it is called the GF 'ATM', we joked). In the proposal, you never have to say how your programme will cut down the likelihood that someone with HIV will pass on body fluids to someone who is not infected, so you can't be held to it. And the Fund doesn't have any way of holding you to it, anyway. It is the biggest health financing experiment in history, and there is no technical oversight. They do send in auditors like Price Waterhouse Coopers to check that you've chatted with 400 drug users, and the auditors make their findings public. If you can tick the 'we spoke to 400 junkies' box, you get a good scorecard. And if you get a good scorecard, so does the Global Fund.*

A few countries, such as the United States, view the Fund with some scepticism. They continue to be the single biggest donor in real terms, contributing US$2 billion of the US$8.1 billion the Global Fund had raised by the end of August 2007, but have hung their contribution about with limits and restrictions. To criticize the Fund is to disagree with the principle of better coordination in global AIDS spending; or to doubt whether individual countries will make good decisions or even really want to fund the things that will make most difference in their epidemic; or to worry that money will be spent on things you don't approve of; or to question the competence of the Global Fund's technical review panel, or all of the above. Even the Bush administration does not dare do all of these things openly, although it does them implicitly by spending 93 per cent of its AIDS money bilaterally.

In some quarters, spending money is a measure of success in its own right. Several years ago, I watched a guy I worked for put up a PowerPoint slide that announced our new corporate mission: 'To increase the burn rate in every country in Asia.'

The burn rate. The rate at which we burn taxpayers' money. Having spent so much time trying to draw attention to the HIV epidemic, trying to get people to take it seriously, trying to get

* The Global Fund did get a bit shirty when they found that the Indonesian government's head of communicable disease control had allocated around US$10 million from a second massive grant to an NGO run by her husband. That's against the rules. All funding to Indonesia was suspended in March 2007, pending an investigation (see Jakarta Post 'HIV projects brush off funding freeze' 11/06/07). The suspension was understandable, even admirable, but extraordinarily inconvenient for groups outside the government that relied on the funds to deliver services. Changes were made to staffing and oversight mechanisms, and funding was resumed after some six months.

money to do what needs to be done, our goal is now to burn money? Apparently, it is a concept with some currency. I once heard a World Bank representative characterize a catastrophic US$44 million loan to Kenya's STI programme as 'a successful project with hardly any impact'. If it had no impact, how was it successful, exactly? 'In the latter part of the project, the disbursement schedule improved.'[23]

We can do better than burn money. Usually we do, and when we don't we have plenty of data to help us understand where we're going wrong. Are we reaching the right people? Are we reaching enough of them? Do the people we reach have less risk than the people we don't reach? Where are most new infections coming from? What are we actually delivering for your tax dollar? By 2004, lots of countries in Asia could answer all of these questions with the information they already had. We could use that information to figure out what we were doing right and what we needed to change. But as I went around sticking my nose into my colleagues' business, I found that it was a dangerous game. Looking for ways to improve programmes implies finding fault with what you're doing right now. People who are busting a gut to do something, anything, for injectors who have zero access to any kind of service don't want some number-cruncher like me to come in and say, 'I know you're working really hard, but what you're doing is making no difference at all to the spread of HIV.' Institutions that want more sugar don't want to show information like that to the donors who control access to the sugar-bowls. And the donors don't want to hear that the programmes they are funding are not doing well, either. So we either report virtually meaningless indicators demanded by the UN or donors, or we torture the statistics to come up with a 'good news angle'.

If you torture the statistics enough, they'll confess to anything.

Doing honest analysis that would lead to programme improvement is a glorious way to be hated by just about everybody.[24] Well, not quite everybody. In the middle of 2004 I sent my regional boss Jeanine a proposal to train people to actually use the information we had to improve HIV prevention services. I gave this radical idea a radical title: 'Data analysis and its use.'

Jeanine hated it. Not the proposal, she loved that. What she hated was my title. 'The terms "data analysis and its use" seem too bland to inspire us,' she declared, and you couldn't fault her for accuracy. As an alternative, Jeanine proposed a question that ought to inform the corporate mission of everyone working in HIV prevention and care:

'What the hell difference are we making, anyway?'

The ten-billion-dollar-a-year question.

9

Full Circle

What the hell difference are we making, anyway? The question preoccupied me more and more.

In the decade since my accidental introduction to epidemiology, I'd learned quite a bit about the things that drive HIV and the things that can stop it. But I still wasn't sure that we were turning what we knew into a better life for Lenny, Nana or Ling Ling, or for the millions of other people whose well-being was the raison d'être for our work. It's not as if we weren't trying. Most of my colleagues were putting in a seven-day week. I'd regularly send off e-mails at midnight so that they'd be in someone's inbox in the morning, only to have them shoot a response straight back at me – they, too, were still at their desks. We were all travelling constantly. I think my all-time air-miles record for a single trip was in January 2005: Jakarta–Los Angeles–Cuernavaca (in Mexico)–New York–Bangkok–Beijing–Jakarta (but only to the airport; a friend brought me some clean clothes, we switched suitcases like minor gangsters in some mafia movie and he took the dirty laundry home for me) then on to Wamena, Enarotali, Timika and Jayapura (all in Indonesian Papua) before heading home to Jakarta. Eleven cities and 256 hours on the work sheet in a month, against a pay slip that will say 160 hours worked. And I wondered why I didn't have much of a social life. There is absolutely nothing glamorous about this sort of travel. It is a mind-numbing succession of alarms going off before dawn, of panicking because you've forgotten your visa, your underwear, the power cable for your laptop. It's waiting for delayed flights, making small talk with faceless bureaucrats, wrestling with recalcitrant photocopiers. And I'm not exactly travelling first class. I spent the flight from Wamena to Enarotali squatting on top of a box of condoms because the plane's three passenger seats were all full.

After a decade of hard work, what was there to show for it?

There were some real successes. At least we now had decent data collection systems that could tell us what was going on in much of Asia. We were getting much better at identifying emerging problems if we wanted to; we now had the tools to figure out where our programmes were falling short. At FHI, we did manage to spend the bulk of the money entrusted to us by USAID, DfID, the Global Fund and others working with the people who really need HIV prevention services – drug injectors, waria, rent boys, prostitutes and guys who buy sex. We were even beginning to work around the edges of a prison programme.

We had some successes with classic Big Tobacco-type lobbying tactics. When we sent a journalist from Indonesia's most influential newspaper *Kompas* to Australia to look at harm reduction programmes for drug injectors, we got three full pages on why it made sense to help junkies avoid HIV infection, and how it could be done. If you tot up the column inches, that's US$47,000 worth of free advertising for harm reduction. In response to the article, the Minister of Health said that Indonesia planned to adopt harm reduction as a national policy. This shocked us all – the merits of harm reduction were still being hotly debated in government. But once it was out of his mouth and on the record, we could start pressing for it to become a reality.[1]

That was a golden moment. But we had to set against that an inordinate amount of time spent finding ways to get around rules imposed by politicians on the other side of the world. We'd have to count things that didn't need counting, and answer questions that made no sense. We banged our heads constantly against the bureaucracies of the countries we worked in, too. Just when you'd convinced some minister that the country desperately needed to stamp on HIV transmission in prison, there would be an election, or a coup, or a death, the minister would change and you'd be back to square one. Most annoying of all, we'd do battle with other organizations which professed to be working towards the same goals, because we both wanted to put our logos on a set of guidelines or be able to report that we'd sponsored this or that initiative. The health ministry would let it happen, because they'd get paid by two sets of donors for doing virtually the same thing.[2] They'd allow different donors to hold separate meetings on the same topic within days of one another and then send the same people to both meetings, because with every meeting comes travel money, per diems and payment for the key speakers. Each of the donors would go ahead with 'their' meeting anyway, because they both needed to tick boxes on their performance scorecard. These things happen in all types of health and development work. They happen more often in HIV programmes simply because there is more ideology, more money, more ants in the sugar-bowl.

For all my whining, I loved my job. I loved working with people like Lenny who opened my eyes to what you can do with a bit of money and a lot of determination. I was inspired by the dedication of Naning Iswandono, my counterpart in the Ministry of Health, though I sometimes felt sorry for her kids. I'd often find Naning in her office at eight or nine at night, four or five hours after most civil servants had shut down their computer solitaire games and fled their desks. Naning did not allow foreign advisers to dictate to Indonesia. She told me when she thought I was suggesting surveillance systems that weren't needed or wouldn't work, when she thought I was being a bully, and she was usually right on both counts. I was always glad to spend time with the aptly named Happy Hardjo – the smiliest man I have ever met – and his colleague Arizal Ahnaf, my mentors in the Bureau of Statistics. Besides teaching me virtually everything I know about running large, complex survey systems, they were fun to be around, mucking in with their staff in a way that was almost unheard of among the Poobahs of the Indonesian civil service. I never got bored or played Bullshit Bingo in a meeting at the stats bureau; we were too busy arguing about what words to use for insertive or receptive anal sex, how we should define 'married' or whether prostitutes would rather have lipstick or nail varnish as a thank you for answering our questions.

But the more information we got, the more frustrating it was that we weren't using it well. I'd suspected we could do more with our data since I first studied epidemiology. I had convinced myself that if I learned the tools of the epidemiologist's trade and brought in some lobbying skills, I could help change the 'Collect data, Publish, Get ignored' modus operandi of public health. I'd done my best. I learned how to run complex epidemiological models, and then present the results under simple headlines. An example was the work I mentioned earlier, showing that effective prevention for drug users was the key to controlling HIV in Indonesia. We'd known how fast HIV can spread through drug-using communities since the early 1980s, but when heroin began to flood into Indonesia in the late 1990s politicians and international health types both looked the other way. I made a graph showing all the infections we'd failed to prevent, and presented it to Indonesian policy-makers in 2003 under a headline which screamed: 'In Jakarta we started late! If we had prevented an epidemic among IDUs from the start, there would have been virtually no HIV epidemic in Jakarta by 2010.'

The presentation drew an explosion of frustration from Chris Green, who in 1995 cofounded the Spiritia Foundation, Indonesia's first peer support group for people with HIV. Chris has been a tireless promoter of sensible HIV policy ever since. 'We didn't start late . . . we started in 1998,' Chris wrote to me. 'But no one listened . . . We talked, we cajoled, we became hoarse, and we were taken for fools. Talk about voices crying in the wilderness . . . You will say, "Yes, but now we have the evidence". We had all the evidence in the world then – literally the world! But no, they said, Indonesia is different.' Chris was beginning to think that repeating other people's mistakes was part of the human condition. 'We are congenitally incapable of believing until it is stark in front of our faces . . . if we are to succeed, it's not just an early-warning system we need, it's some form of "cattle prod" to galvanize a reaction.'[3]

Here we were, halfway through the first decade of the twenty-first century, and we were still looking for cattle prods, even in places where the death and destruction was stark in front of our faces. The language of the endless reports on AIDS became ever more apocalyptic. In January 2005 I got a despairing e-mail from Keith Hansen, who managed the World Bank's AIDS portfolio in Africa. He'd just been given another 'expert' paper on AIDS to comment on. 'It should be called, "The Day the Moon-sized Meteor Strikes Earth"' he wrote to me wearily. 'Not a single new word or shred of data, not a single insight nor constructive suggestion. Just rehashed, undifferentiated fear-mongering. The only people who are not identified as "especially vulnerable" are eunuchs and the dead. Oh, and drug users, of course, since they don't exist . . . I know the sky is falling, but simply repeating it time and again won't prop it up.'

We'd come a long way from the excitement and optimism that pulsed through the corridors of UNAIDS in its earliest days. Nowadays, the dominant sensation was déjà vu. After ten years as an epidemiologist, I was still writing screeds on the better use of data, still looking for money to train people in 'Data analysis and its use', still singing the same old tune. I was starting to bore *myself* with the refrain, so God knows what I was doing to my poor colleagues.

One day, I found out. It was right in the middle of that eleven-city trip-from-hell, and I was in Bangkok for a meeting about cattle prods – tools we could use to encourage politicians to do the right thing to prevent HIV in their particular epidemic. Two of my favourite people were there – John Stover, a computer modelling nerd whose buttoned-down exterior hides a warm sense of humour and an unlikely past as an organizer of traditional weavers in Uruguay, and Tim Brown, an astrophysicist turned epidemiologist who subsists on a diet of Pepsi Max and endorphins.

In a meeting with USAID, Tim had been describing his progress in refining the Asian Epidemic Model, the very one I had used to fashion my failed cattle prod for Indonesian policy-makers. The model calculates what happens to HIV when injectors buy sex from women, when married men sell it to men, and when people do all of the other wonderful, fun and stupid things that put them in the way of HIV in Asia. He was throwing up equations like this:

$$[\mathrm{CstdFstd}(t) + (1 - \mathrm{Fstd}(t))]\ [\mathrm{CccFcc}(t) + (1 - \mathrm{Fcc}(t))]\ \mathrm{Pf\_mX1V1}(t)\ (1 - \mathrm{C1}(t))\ [\mathrm{Y3} / (\mathrm{X3} + \mathrm{Y3})]$$*

We all nodded as if we had any idea what he was talking about.

Then it was John's turn. He was explaining a model called Goals, which aims to help policy-makers calculate how much they need to spend and how they need to spend it to get the most bang for their buck. It is a great idea, but John had been working mostly in Africa, and the version he was showing us was never going to prod Asian politicians in the direction they most needed to go – towards the needle exchange.

Wearily, I raised my hand and interrupted John's presentation. 'Just for the record . . .' and I began to tick off on my fingers all the reasons that we shouldn't export African tools to Asia. a) . . . b) . . . c) . . . d) . . . e) . . .

John heard me out, in his usual buttoned-down polite way. Mmmm. Yes. Uhuh. Then, when I was quite finished, he paused, said 'Because I knew Elizabeth Pisani would be here today . . .' and clicked on to the next slide.

It was headlined 'Disclaimers'. And there were all my objections. a) . . . b) . . . c) . . . d) . . . e) . . ., even in the same *order* as I had listed them, goddamnit!

When the meeting was over, I walked across the hall to Jeanine's office and quit my job. It's bad enough that you bore

* Just in case you're curious, this formula is for calculating the likelihood that a client will get HIV from a sex worker. It translates as: increase in risk if you have an STI, plus infection rates without STIs, plus increased risk if you're not circumcised, plus infection rates in the circumcised, all times the probability of HIV transmission per act of sex from a woman to a man, times the likelihood that sex is unprotected, times the total number of sex acts, times HIV prevalence among sex workers.

yourself with your refrain, but when you've been singing the same song for so long that other people know the words by heart, it's time to let someone else have a turn at the mike.

Magic bullets

HIV has changed the landscape of public health. It wasn't doctors who kick-started the response to this infectious disease, it was gay men. With a flair for dramatic presentation and an inside knowledge of what makes the communications industry tick, they battered down the doors of the medical establishment, assaulted the pharmaceutical industry, took the press by storm. They gave AIDS a political face. They were the production team that made AIDS different from the competing acts, that got AIDS started on its journey up the international development hit parade.

AIDS has rewritten the record books in terms of cash for an infectious disease. We should give George Bush and his conservative supporters credit for getting the money on the table. It was they who raised the stakes for everyone else in the game, they who shamed the world into taking AIDS treatment for poor people out of the 'too hard for now' basket and into the realm of possibility. Because everyone wants to meet their compassion targets, donors are being forced to take on issues like chronic corruption, inefficiency and understaffing in the health sector – the quicksand that swallows up development efforts but about which we've done pitifully little for decades.

The fashion for investing in AIDS has brought new players into the field of international public health: the Global Fund, the New Philanthropists, Bono. Not wanting to be outdone by the upstarts, rich countries doubled their spending on health in poor countries from US$7.2 billion to US$15.7 billion between 2001 and 2005. AIDS has sparked new ways of financing medical research for diseases that affect people in poor countries, prompting a more international focus at foundations like the Wellcome Trust, and the creation of public–private partnerships such as the International AIDS Vaccine Initiative. People with diseases that are not covered by high-paying medical insurance schemes have been released from the tyranny of Big Pharma's shareholders. Where AIDS slashed a trail, malaria and TB have followed. Organizations like the Global Fund are setting new standards for transparency in doling out money. They are focusing far more on bean-counting than they are on asking whether a country really needed beans in the first place, but at least they are counting, and that's a start.

A lot of the money still goes where the biggest donors say it should, and that is not always to the right place. But AIDS has also broken ground on a more collaborative way of working. Donors now at least pay lip service to the needs and desires of the people they are supposed to be trying to help, and their governments. For better or for worse (and usually for better, I'd say) AIDS has started to undermine the non-consultative 'do as I say' culture that used to be the norm in much development work.

Does that mean the money is more likely to go where it should? It rather depends on whether a government wants to do what people in a country need. The record is not encouraging on that score; if it were, we wouldn't be in this mess in the first place.

Some of things that HIV prevention money gets spent on are helpful in their own right, even when they don't actually prevent any HIV. Sex education is useful to kids even when they're not at risk of contracting HIV – talking openly about sex and condoms at least cuts down on unwanted pregnancies and STIs. If you improve the safety of the blood supply, you avoid hepatitis and other viruses as well as HIV; it's a good thing to do even when there's no HIV around.

But the downside of programmes that focus on people who aren't at risk is far greater than the upside. Firstly, doing the wrong thing eats up the time and energy of people who are already short of both. An East Timorese health ministry official who is in Geneva for a stakeholders' consultation on community responses to HIV is an official who is not back home trying to work out how to cut down the number of women dying in childbirth, or the number of children dying of diarrhoea. Secondly, HIV prevention programmes that don't focus on reducing the likelihood that infected people will pass the virus on to uninfected people make governments, voters and even people who buy Bono's red iPods feel like they are tackling the HIV epidemic when in fact they are completely missing the plot.

In the early years of the epidemic this was excusable – we really weren't too sure where things were headed. Now it is beyond inexcusable – it verges on the iniquitous. It is as if we can see a train coming down the track and we know it's headed for a crash. Perhaps one of those commuter trains you see in Jakarta, packed to the gills, with people hanging out the doors, standing between carriages, sitting on the roof. All we have to do is throw a couple of switches, change the points, and the train would trundle off on a safe path. No crash. No makeshift morgue with shattered corpses in sticky crimson rows.

We know where we are headed with HIV. We've seen the train coming in one country after another, and we know which switches we have to flick to stop the crash. But a lot of the time, we're not doing it. We just look at the train and say, 'What are all those people doing sitting on the top of the train anyway? That's illegal. And stupid.'

We don't do it because no one wants to look too hard at the truth about HIV. We don't tell the truth for fear of seeming racist, for fear of losing our jobs or our chance for a promotion to a director's position, for fear of seeing our institution's budget evaporate. We don't tell the truth for fear of upsetting people who are already infected with HIV, or stigmatizing people who belong to groups in which HIV rates are high. We don't tell the truth for fear of losing clients, access to health care, our marriage. We don't tell the truth because our religions and our cultures want us to be prudish about sex and drugs, whereas, in truth, most of us think they are fun.

But we can't solve a problem that we won't describe honestly.

If you've got this far in the book, you'll have a pretty good idea of what I think the problem is. But let me try and summarize.

The Problem (on Planet Epidemiology)

- The overwhelming majority of HIV-infected adults get HIV by having unprotected sex with an infected person, or by sharing needles with them while injecting drugs.
- People pass the virus on most easily when they have lots of it in their blood.
- HIV is transmitted most easily when it comes into contact with sores, lesions or a foreskin.

The Solution (on Planet Epidemiology)

- **Cut the exchange of body fluids between infected and uninfected people.**
  Getting people to give up sex doesn't work very well, so in terms of sexual transmission, our best bet is to persuade

uninfected people to use condoms with any partner likely to be infected with HIV. In East and Southern Africa and in many gay communities that means any new sex partner. Bombard the places people go to meet new partners with condoms and lube. Remember that men have sex with one another in prison.

For heterosexuals in most of the world, the highest-risk sex is paid for. Create incentives to use condoms every time sex is bought or sold. Use blackmail and bribery if you need to.

Methadone and other oral drugs can help people stop injecting, so make them widely available but recognize that they don't always work. Make it as easy as possible for people who are still injecting to do it safely. Make clean needles available, make them free or dirt cheap. Give them to people in prison if they need them. Make sure no one winds up in prison just because they're carrying a needle.

Don't forget to make sure that all injecting equipment used in health services is sterile. Screen all blood for HIV before transfusion.

- **Cut the amount of virus in the blood of people who are infected**. The most obvious way to do this is to provide people who need it with antiretroviral treatment. Test for infection where you're most likely to find it. Bundle prevention services with AIDS drugs at every opportunity. Treating other infections (especially sexually transmitted infections) promptly will reduce spikes in the amount of virus, even among people who are not on anti-HIV drugs.

If you are infected and you're pregnant, the best way of keeping your child HIV-free is to keep your viral load down around childbirth. Antiretrovirals are at the core of programmes to prevent HIV passing to infants. Good infant formula milk is important too, because it replaces infected breastmilk.

- **Close all the potential 'open doors', so the virus can't get into the uninfected person.**
  In communities where there is lots of HIV, circumcise men. Screen for sexually transmitted infections and treat them among women, men and transgenders who sell sex, and among their clients and regular partners. Make sure you use drugs that work. Promote the use of lubricant in all anal sex and for female sex workers, too, to minimise tears and lesions. Package it with condoms. Get it into gay bars.

It's all pretty straightforward, and that begs one obvious question: if that's all there is to it, why haven't we done it yet? You might have a bonus question, too. If that's all we have to do right across the globe, why the big deal about understanding the local landscapes of sex and drugs?

I'll deal with the bonus question first, since it partly answers the more difficult earlier question. The list above tells us *what* we have to do. It is determined by biology. There are only a limited number of ways HIV can get from one human being to another, and so there are only a limited number of ways we can prevent that happening. Those are the *what* questions. But the list doesn't tell us *how* to do it, or even exactly *who* to do it for. That will be determined by the local landscape. A Thai-style '100 per cent condom programme' that relies on blackmailing brothel owners isn't going to work in Laos, where most sex workers freelance on the streets. The *what* is the same: you still want guys to use condoms every time they buy sex. But the *how* will be different. The *who* may vary as well, according to the local topography. We want all drug injectors to be able to use a clean needle for every fix. But what if you find out that 30 per cent of the girls that shoot up are selling sex? To prevent the greatest number of HIV infections you need to make sure they get clean needles before anyone else, because they're more likely to pass the virus on than an injector who just has sex with his girlfriend.

If you do HIV and behavioural surveillance in the right groups and you look at the data with an open mind, your results will lead you to the *who*. But it is very often the '*who*'s themselves that will lead you to the *how*. This is not a piece of 'peer outreach' orthodoxy. It is simply a matter of experience and common sense. I can tell Lenny exactly which behaviours have to change to bring down HIV transmission among waria. They need to try and shift more clients to oral sex (difficult, because most waria prefer anal sex and get paid more for it), use more condoms and lube in anal sex, get screened and treated for anal and urethral STIs. Now that antiretrovirals are beginning to become available, they need to go for HIV testing so that they can get treatment if they need it. None of this is news to Lenny. But some of what she and her friends tell me is news to me. After holding five discussion groups to review the results of the study (one group for each administrative area of Jakarta), the assembled waria decided they wanted an ugly male doctor for their STI clinic. They didn't want a woman messing around with their genitals. But they didn't want a good-looking man, either, because they were worried they would get erections while he was examining them, and wouldn't that be embarrassing. Ignore these details at your peril.*

* In the end, apparently because someone at FHI wanted to give a job to a friend, the waria clinic got a female doctor. Clinic attendance has been lower than was hoped for.

The *what* may be simple and the *who* is epidemiology, not rocket science. But the *how* can differ quite a bit from place to place. It usually takes time and patience to discover, and programmes driven by quarterly indicator reports to donors don't always have the time or the patience. That's part of the reason why we haven't yet done all the things we should have done, the simple things that would put a brake on the HIV epidemic.

I confess that many of us on Planet Epidemiology tend to gloss over, even ignore, a few of the handful of things that are biologically effective in preventing HIV. The first is to stop people who are HIV negative from having any sex with people who may be infected: abstinence. It works for the people who stick to it, but it has a high failure rate. The second is to get people to shift from having several partners at once to having just one at a time (this reduces the likelihood that an uninfected person will have sex with someone who has only just picked up HIV, whose body fluids will be loaded with virus to pass on). We've seen evidence of this in one country, Uganda. Cutting down on girlfriends and boyfriends hasn't caught on elsewhere yet, and in Uganda there are signs that people are partnering up again. There's more evidence that men are prepared to cut down on getting drunk and tipping off to the brothel – commercial sex has fallen dramatically in Thailand, Cambodia and parts of India, but it has held fast in many more places. We downplay these potential 'solutions' on Planet Epidemiology because we don't think they work very well in the Real World. We call our reservations evidence-based, though others would say the public health establishment is being ideological in its own way.

This goes right back to my opening observation that science doesn't happen in a vacuum. Scientists can identify problems and solutions any way they please, but scientists almost never get to impose solutions in the Real World. People who call the shots in that world come from a different planet, where even the problem is different.

The Problem (on Planet Politics)

- In East and Southern Africa, most adults get infected by having sex with someone who is not a single, lifelong partner. In the other 97 per cent of the world, the overwhelming majority of HIV-infected adults get HIV by shooting up drugs, having anal sex, buying or selling sex.
- The overwhelming majority of voters have been told by their religious leaders, their elders or their politicians that these behaviours are wicked. So even if they do them, they rarely demand help to do them more safely.
- There are not many votes in doing nice things for wicked people. Rich countries don't want to pay for it and poor countries don't want to do it.
- Recognizing this, the AIDS mafia tried to turn HIV into 'everyone's problem', a development problem, a security problem. True in 3 per cent of the world. Untrue everywhere else. But rich countries will pay for security problems, so poor countries can get money to do nice things for people who are not wicked. There are votes in doing nice things for people who are not wicked.

The Solution (on Planet Politics)

- Weigh up the options. Do I scrape around for money to do things that will lose votes, or get showered with money to do things that will win votes? Hmmmmm . . .

To a politician in any country outside of East and Southern Africa, that's a pretty straightforward choice. Many of them won't even know that the things that win votes will do nothing to slow the spread of HIV. They've been subjected for a very long time to the notion that HIV is a disease of poverty, gender, fisheries. Some will know that if they used the money properly they could save the lives of millions of drug injectors, of women, men and transgenders who sell sex, of men who buy it, of men who have sex with one another. But politicians who know this may well not care.

In the countries where HIV is highest, your calculus should be different. The virus is spread by something that most people do, even if they are told that it is 'wicked'. If everyone recognized exactly what it was that was spreading HIV – sex in nets rather than strings, older men sleeping with younger women, untreated STIs, too many foreskins, a willingness to risk death for the added pleasure of sex without a condom – societies could rearrange themselves to deal with it. President Yoweri Museveni of Uganda has not been thrown out of office for talking about AIDS, sex and condoms, for dealing effectively with the HIV epidemic. On the other hand, South African President Thabo Mbeki hasn't yet been thrown out of office either, and he fiddles with the truth while the virus burns through his nation. Even where most of your voters are at risk, you can apparently get away with ignoring HIV on Planet Politics.

I chose a career in sex and drugs because I believed it ought to be possible to convince politicians that the epidemiologists' solution to the HIV epidemic would make political sense too. It started out as an intellectual exercise. But as I spent more time with people in the firing line, people like Lenny and Ines, Desi and Ling Ling, Al and Adrian and Chris, my interest morphed. It's almost embarrassing for a cynical, wise-cracking, number-crunching ex-journalist to admit, but I began really to give a damn. Getting HIV prevention services for people who needed them most has begun to seem like a debt I owe to these friends and the countless other people who have taught me about sex, drugs and HIV. I am in their debt for the help they've given me and the wisdom they've shared, but also because I feel I contributed to the 'Everyone is at risk' nonsense that is now getting in the way of doing what is needed. When I left Indonesia I felt I had failed my friends and colleagues. It seemed that no amount of good science could overcome bad politics and the power of money to rub out common sense.

In the AIDS industry, I felt, we were all whores.

Too much of a good thing

The most recent chapter in my AIDS adventures brought me full circle. My very first conversation with a prostitute was in a bar in Hong Kong in the late 1970s. Since then, the British colony has been grafted back on to China, and it is there that I've learned my latest lesson. Many of Hong Kong's strengths have seeped back into the body of the motherland since they were reunited in 1997. A vibrant financial services industry, for one thing. And a vibrant sex industry, too. Communist apparatchiks don't need to go to the Club Volvo wedding cake for their entertainment any more; Beijing now provides clubs that are just as honey-soaked and a lot more expensive. In late 2005 I found myself in Beijing, chatting with prostitutes and helping the said apparatchiks figure out what was going on in their HIV epidemic.

The wise old men of the Chinese politburo had closed their

eyes to HIV throughout the 1990s, to the frustration of the handful of people in the health ministry who were trying to shake them awake. They were eventually roused from their torpor by the blood transfusion fiasco I described earlier. In 2001, after news of the dirty blood scandal broke, China's leaders instructed the health ministry to find out what was really going on. Ministry staff worked with a few experts from Chinese universities to forecast the epidemic, plotting reported HIV cases on a graph and drawing a straight line through them. This is so many different kinds of wrong it's not even worth explaining why. The method resulted in a ludicrous prediction that China was headed for 20 million HIV infections by 2010. The ministry sprang this on UNAIDS epi-nerd Neff Walker during a meeting in Beijing in late 2001, looking for a rubber stamp of approval. Neff was horrified, and did what he could to damp down the absurd numbers without offending anyone. With no preparation time and no decent data to work with, he shuffled a few numbers into a spreadsheet and came up with an on-the-spot guess of the current situation: up to 1.5 million people infected.[4] And the worst that could happen in China by 2010? If HIV prevalence among all sex workers in China reached the same levels as they did at the height of the epidemic in Thailand, *and* all gay guys in China got as infected as gays who trawl the bars of San Francisco, *and* everyone who injected drugs was as likely to be infected as a junkie in Burma, then you might get to 10 million. 'It may all be crap, as I still have little belief that this approach really captures much,' he wrote to me at the time. Even if the 10 million figure was crap, it was only half as crap as the experts' estimate of 20 million. But it was never meant to be a prediction of what was likely to happen in China.

None of this mattered to the UNAIDS office in Beijing,

which seized on the figure with unmitigated glee, featuring it in a 92-page doom-fest entitled 'China's Titanic Peril'.*[5] Orchestrated to prod a reluctant Chinese government into action, it was launched with loud drumrolls and great clashing of cymbals.

The hysteria prompted the Chinese government to try and make better estimates. Wisely, they concentrated just on the here and now, rather than trying to predict the future. In 2003 they took Neff's spreadsheet, filled in some of the gaps with data rather than guesses and came up with a figure of between 650,000 and 1.02 million people infected with HIV in China. 'Cover-up!' yelled AIDS activists around the globe, assuming that the earlier UN figures must be more accurate that anything the Chinese government came up with. Journalists followed suit. 'You can't blame us,' *The Economist*'s Beijing correspondent told me several years later. 'We all had that 10 million figure burned into our brains.'

All the fuss seems to have contributed to breaking the political dam. Suddenly, in 2004, AIDS shot up the agenda of the old men on the politburo. Bigwigs visited AIDS patients for the TV cameras, and the Four Frees and One Care programme I described in Chapter 5 – the free services and schooling that people were at first so reluctant to sign up for – was committed to. But Beijing was stung by accusations of number-squashing. So in 2005 the Chinese government launched a massive, nation-wide programme to try and document what was really going on with the virus.

* The 'Titanic' report attributed the 10 million figure to the 1998 Mid–Long Term Plan of HIV/AIDS Prevention and Control in China (1998–2010), State Council, 12 November 1998, thus implying that it is an official government figure. However, I can find no trace of any projection beyond 2000 in that document.

I'd been to China a couple of times since 2003 to help pass on to health ministry staff some of the lessons we'd learned while estimating the number of drug injectors, sex workers, clients and others at high risk of catching HIV in Indonesia, as well as the numbers already infected. After I left my job at FHI, the Chinese government asked if I'd lend a hand with the new round of estimates. UNAIDS would pick up the tab. I really liked the crew of bright young things in the epidemiology division of China's national AIDS programme, so in the autumn of 2005 I agreed to sit with them in their dreary 1960s back-lot offices and crunch numbers. I was billed as a 'foreign expert' but in truth my boss Lu Fan and his team didn't really need my expertise – they were hugely competent. I'd occasionally manage to throw in some scrap of wizardry from the estimation toolbox, but we all knew what my real role was. Just by sitting there, I protected my Chinese colleagues from any pressure from Chinese politicians to cook the books ('We'd love to bash down the numbers but there's this girl in the office . . .'), while at the same time protecting the politicians from being accused of a cover-up ('New estimates of HIV in China, made in collaboration with UNAIDS experts . . .'). The political higher-ups appeared really to want to know the truth. They charged us to come up with a new estimate that they could publish on 1 December, World AIDS Day, and we worked fiendishly to do it.

A lot of the time I just sat around getting in the way of my Chinese colleagues and pretending not to notice the smell of stale urine which seems to hang over all government offices in China. My friends were very patient with me as I tried to rub the rust off my Chinese and muck in as best I could. We travelled to all corners of the land, asking local officials to help us understand unexpected blips in the data. 'Oh, the 19 per cent infection in sex workers? Well, it's because the drug rehab centre got flooded, so they put the junkies in with the hookers.' Everything was carefully recorded in hundreds of annotated spreadsheets – two each for almost every prefecture in China. It took months of late nights in the office, of cajoling local officials over banquets, of drinking far more unspeakable white spirits than was good for any of us, of begging the patience of Geneva, which also wanted a number to put in their World AIDS Day report but had a longer lead time for their publications. It took endless meetings and presentations and discussions punctuated by cups of green tea poured by drifting waitresses in white gloves who tried to keep a straight face as greasy officials with bad comb-overs protested: 'So much anal sex? Not in my prefecture!' But by the end, China had estimates for the numbers of prostitutes, clients, drug injectors, gay men and former plasma donors for over 300 prefectures in the country, as well as estimates for the numbers infected with HIV. Every piece of data was sourced, every assumption annotated. There were a few weaknesses; for example, infections in gay men were tipped upwards by Beijing-based staff for reasons I was never able to fathom. But overall, Lu Fan and his colleagues had produced possibly the best HIV estimates that we'd seen in any developing country. And they'd done it on deadline, so that the political higher-ups could do their drum-rolling on World AIDS Day as promised.

The political higher-ups took one look at the new estimates and decided not to go public. We were shocked. This was post-SARS; the Ministry of Health had assured everyone that they were committed to telling the truth, that they would follow wherever the data led. We'd even held mock press conferences to prepare for the release of the results (I played the grumpy foreign correspondent). And now, after all that hard work, after arriving at a number that I truly believe was the best possible estimate for China, the politicians were reverting to type. If you don't like the truth, cover it up.

The problem was: the numbers were too low.

From a top estimate of 1.5 million infections in 2001 in the 'Titanic' report, the estimates for 2005 had just about halved to a high of 760,000. A couple of years earlier, China's top officials would have been thrilled with an internationally sanctioned figure as low as this. But by the end of 2005, times had changed. The country's leaders had made a big deal of their commitment to fighting AIDS. They had applied to the Global Fund for US$205 million in funding for the problem (and been granted US$180 million). And they had the likes of the Bill and Melinda Gates Foundation queuing up to shove another 50 or 60 million bucks their way. The government was investing massive amounts of its own money in programmes for migrants, pregnant women and other groups that weren't at much risk of HIV infection in China. They had invested a lot of 'leadership' in AIDS. To shrink the problem now was a loss of political face.

To their credit, the Chinese government did eventually publish the new estimates, in a joint report with UNAIDS and WHO.* To their even greater credit, Beijing listened to the public health specialists such as Wu Zunyou, who heads the AIDS division in the Ministry of Health, and refocused their prevention efforts on drug injectors, gay men and the sex industry.[6]

* In an extraordinarily clever piece of media management, the Chinese government allowed the feeding frenzy of World AIDS Day to pass without issuing any new estimates. Word of the new estimates leaked out slowly through 'technical briefings' which swamped people with nerdy details about data sources and methodologies. By the time the new estimates were released on 25 January 2006, no one had the energy to be surprised, let alone to worry about an imagined cover-up. For a description of the estimates, see Lu et al., 2006.

What happened in China restored some of my optimism. After years of denial, the most populous nation in the world is making sensible decisions based on a sensible reading of well-documented, scientific estimates of who is infected and who is most at risk. I find this hugely encouraging. Some poor countries still bury their HIV epidemics under a blanket of silence and beetroot. Some rich countries shackle their funding for AIDS in bonds of ideology that no amount of science can cut through. Too many institutions still hold fast to the 'everyone is at risk' mantra so that they can keep shovelling sugar out of the bowl of AIDS funding. Millions of people go on doing things they know are risky because they care more about pleasure now than AIDS in some unimaginably distant future. I'm one of them, and I'll guess many readers of this book are, too.

Together these failures have added up to 70 million HIV infections so far. Just writing that figure will date this book, for it is bound to grow. But slowly, very slowly, more and more countries are following where Uganda and Senegal, Thailand, Cambodia and Brazil first led. They are being chivvied along by Britain and the Netherlands, by Sweden and Ireland, by a growing number of countries that think that science – rather than ideology, religion or the self-interest of bloated institutions – should dictate what we do about HIV.

In the last two decades, people like Lenny and Nana and Al, people like Naning and Happy and Jeanine, people like me and maybe people like you, have become footsoldiers and spear carriers in a war whose weapons have been ideology and religion and culture, hypocrisy, shame and prejudice. But there's another arsenal: we have the knowledge, the tools and the money to beat the HIV, and many other diseases besides. The armies on Planet Epidemiology and Planet Politics have been waging their battles in isolation from one another, the one fighting for things that are effective but unpopular such as clean needles, the other fighting for things which are ineffective but popular such as abstinence. But we are beginning to see signs that common sense and common humanity are gaining the upper hand. If politics and epidemiology got off their separate planets and stood shoulder to shoulder in the war against disease, we'd have a better Planet Earth.

NOTES

Preface: The Accidental Epidemiologist

1 The legendary paper which began the anti-smoking movement was published in 1950. (Doll and Hill, 1950). The Surgeon General's first report on the subject was published on 11 January 1964. (US Surgeon General, 1964).

2 See the 2001 editorial 'Our Policy on Policy', *Epidemiology*, 12 (4), 371–2.

3 Reagan mentioned AIDS for the first time in response to a reporter's questions on 17 September 1985. For a transcript, see http://www.aegis.com/topics/timeline/RReagan-091785.html#R85.

4 For a detailed history of the AIDS epidemic in the UK, see www.avert.org/uk-aids-history.htm.

Chapter 1: Cooking Up an Epidemic

1 Brown, 1992; Merson, 2006.

2 See for example http://ghiqc.usaid.gov/tasc3/docs/442-07-001/rftop_442-07-001.pdf.

3 E-mail 04/10/02. The document was eventually published as 'Estimating the Size of Populations at Risk for HIV: Methods and Issues', *Family Health International*, 2002.

4 You can find the results of their deliberations at www.data.unaids.org/pub//InformationNote/2006/EditorsNotes_en.pdf.

5 Data for 2006 from the American Society of Plastic Surgeons, www.plasticsurgery.org/media/statistics/loader.cfm?url=/commonspot/security/getfile.cfm&PageID=23761. Accessed 27 July 2007.

6 UNAIDS and WHO, 1996.

7 UNAIDS and WHO, 1997.

8 UNAIDS, 1998c.

9 Chin, 2007.

10 The methods are described in detail in Schwartlander et al., 1999. The model was developed by Jim Chin, who subsequently led the 'UNAIDS is making it up' chorus. See reference 9.

Chapter 1: Cooking Up an Epidemic (continued)

11 Burton and Mertens, 1998.
12 UNAIDS, 1998a.
13 The cost of the report does not appear in the UNAIDS report to its coordinating body. These figures were reported to me in early 2007 by a colleague involved in the 2006 publication. An exact accounting was said to be difficult because there is no clear budget line for the report; several different departments' funds were tapped to pay for it.
14 Lawrence Altman, 'Parts of Africa showing H.I.V. infection in 1 in 4 adults', *New York Times*, 24 June 1998.
15 World Bank, 1997.
16 The World Bank meeting was held in January 1988; its results were summarized in a World Bank advocacy document as well as in four papers collectively published as: 'Demographic Impact of AIDS', Carael, M. and Schwartlander, B. (eds.), AIDS 1998 12: Supplement 1.
17 National Intelligence Council, 2000.
18 USAID's response to the congressional request was submitted in 2000 as 'The United States Responds to the Global AIDS Pandemic: An Analysis of Projected Targets, Goals and Resource Needs'.
19 Attaran and Sachs, 2001; Schwartlander et al., 2001b.
20 UNAIDS, 2006. Unpublished annual breakdowns courtesy of UNAIDS Asia-Pacific.
21 Synergy and USAID, 2001. The contribution of governments to financing HIV prevention in their own countries can be misleading. Much of it is required by donors as 'matching funds', as a way of demonstrating the home government's own commitment to fighting HIV. This contribution is often made not in cash but by calculating a notional value for the use of government offices to hold meetings, etc.
22 Katherine Seelye, 'Helms Puts the Brakes to a Bill Financing AIDS Treatment', *New York Times*, 7 May 1995.
23 Chin and Mann, 1989.

Chapter 2: Landscapes of Desire

1 Linnan, 1992; World Bank, 1996; Elmendorf et al., 2004; Jalal et al., 1994. See www.worldbank.org/ieg/aids/docs/case_studies/hiv_indonesia_case_study.pdf.
2 These estimates were arrived at in a workshop facilitated by WHO consultant Jim Chin in March 2001.
3 Lubis et al., 1997.
4 Hull et al., 1997.
5 For example, 25 per cent of unmarried urban men in a household survey in Cambodia said they had bought sex in the last year,

Chapter 2: Landscapes of Desire (continued)

compared with 18 per cent of married men. In rural areas there was less difference; 15 per cent of the unmarried were clients compared with 13 per cent of married men. Differences in proportions married mean that about half of all clients of prostitutes in Cambodia are married men. Source: Heng et al., 2001.

6 Of over 2,000 men in high-risk groups who told us they had sex with girlfriends, exactly half later said that they had paid those 'girlfriends' cash after sex. Some 11 per cent of them said they used condoms at last sex with their paid girlfriend compared with 30 per cent condom use with partners described as sex workers.

7 Elizabeth Pisani, 'Hong Kong's nightclub glamour is not what it seems', Reuters, 4 May 1988.

8 In some countries, including Indonesia, it is now possible to avoid the cruising scene by using internet chat rooms and small ads in the paper. An example from the *Jakarta Post*, Indonesia's leading English language daily: 'BALI BROWN MASSAGE. Full body. In Bali. For man, by man. Call now Anton [phone number] 24 hours'. Personal ad, 12 December 2006.

9 Behavioural surveillance among 345 male sex workers, 2004. Indonesian Ministry of Health/Statistics Indonesia.

10 Zhang et al., 2007.

11 For example, in Indonesia's easternmost region, more than three-quarters of unmarried teenage Papuan girls who drink reported unprotected sex, compared with just over a quarter of non-drinkers. Some 54 per cent of Papuan boys who drank reported unprotected sex, compared with just 14 per cent of non-drinkers. Nationwide, men who said they used drugs were 50 per cent more likely to buy sex than non-drug users, and drug-using clients were also less likely to use condoms in commercial sex – less than a quarter of drug-using clients said they used a condom the last time they bought sex, compared with close to a third of non-drug takers. Among female prostitutes, 49 per cent of drug users reported condom use, compared with 58 per cent of non-users. In the Indonesian data the only groups for whom behaviour is equally risky regardless of drunkenness or drug use are the groups where those behaviours are most common: gay men and waria. All these data come from the national surveillance system operated by Indonesia's Ministry of Health and Statistics Indonesia.

12 It turns out it is not *every* closet. A 2004 survey conducted by Yayasan Cinta Anak Bangsa, in cooperation with the National Narcotics Board and the University of Indonesia Institute of Applied Psychology, found that just one in ten families in Jakarta has a member involved in drug abuse.

Chapter 2: Landscapes of Desire (continued)

13 Of all the groups included in behavioural surveillance in Indonesia, illegal drug use was highest among self-identified gays, 52 per cent of whom said they were users. Next, however, higher than waria or male and female prostitutes came school kids. In Jakarta in 2002, 34 per cent of boys said they had tried illegal drugs, while just 9 per cent had ever had sex. More than three times as many boys reported injecting drugs as reported sex with more than one person in the last year. Among girls, illegal drug use was slightly more common than sex; 6 per cent admitted to using drugs in the last year, compared with 5 per cent who said they had lost their virginity. Source: Indonesian national surveillance data, Ministry of Health/Statistics Indonesia.

Chapter 3: The Honesty Box

1 Hamilton and Kessler, 2004.
2 Schwartländer et al., 2001a.
3 Family Health International, 2000; Mills et al., 1998.
4 Family Health International, 2002.
5 Indonesia Directorate General of Communicable Disease Control and Environmental Health, 2003.
6 For detailed guidelines on managing data from national surveillance systems, see Family Health International, 2006. Of all the 'cookbooks' I have ever written this is the most tedious, but by far the most useful.
7 World Health Organization, 2001.
8 Monitoring the AIDS Pandemic, Network (MAP), 2001.
9 For a more detailed discussion of the impact of decentralization on public health surveillance in Indonesia, see the briefing paper on www.wisdomofwhores.com/references.
10 Kumalawati et al., 2002.
11 www.fda.gov/foi/warning_letters/m2195n.pdf, accessed 26 April 2002. A copy of this letter is posted at www.wisdomofwhores.com/references.
12 Overall, 2.5 per cent of gay men and 3.6 per cent of male sex workers were HIV infected, compared with 22 per cent of waria. See Pisani et al., 2004 for details.
13 UNAIDS, 2003.

Chapter 4: The Naked Truth

1 We are accustomed to thinking about HIV in terms of 'prevalence', the percentage of adults aged 15–49 infected right now. In a handful of countries, this figure has already topped a third. But it conceals a reality that is even more appalling. Many people die of HIV well before they are 49, so people who would be counted as HIV infected will have

Chapter 4: The Naked Truth (continued)

already dropped out of the statistics. What's more, many people get infected well after 15, so there are people in the cross-sectional percentage measure who are not yet infected, but who will eventually get HIV and die from it. If you add the 'already dead' and the 'not yet infected' to the 'currently infected', the prevalence figure can more than double. See Blacker and Zaba, 1997.

2 HIV prevalence among a nationally representative sample of 15–19-year-old girls in South Africa in 2005 was 9.4 per cent (South Africa Department of Health, 2006). HIV prevalence among female sex workers in China at end 2005: 0.7 per cent (Lu et al., 2006). HIV prevalence among pregnant women aged over 40 in Swaziland, 2004, was 38 per cent. HIV prevalence among pregnant women in this age group underestimates population prevalence because of reduced fertility. In 2003 some 37.5 per cent of 4,183 male employees at a large sugar corporation in Swaziland tested positive for HIV. From these figures, I extrapolate that senior civil servants in Swaziland are certainly more likely to be infected with HIV than Australian drug injectors, whose prevalence was around 1 per cent. (National Centre in HIV Epidemiology and Clinical Research, 2006). Source for Swaziland data: National Emergency Response Council on HIV/AIDS, Government of Swaziland (2006).

3 I cannot attempt in this one chapter to do justice to the complexities of HIV in Africa. For more comprehensive accounts see Barnett and Whiteside, 2006; Epstein, 2007; Hunter, 2003.

4 World Bank, 1997.

5 'Regressions explaining the percentage of the urban population infected with HIV'. From a presentation given by Mead Over to a meeting of the UNAIDS/WHO Reference Group on HIV Estimates, Modelling and Projections, held in Rome, 8–10 October 2000. See Over, 2000.

6 For a reliable and thoroughly documented description of the links between social and economic meltdown, risk behaviour and HIV, see Barnett and Whiteside, 2006.

7 Per capita income ($US): Botswana 5,073. South Africa 4,675. Mali: 371. Niger 228. Female literacy (per cent): Botswana 82. South Africa 81. Mali: 12. Niger 15. Percentage of adults infected with HIV: Botswana 24. South Africa 19. Mali: 2. Niger 1. (Sources: UNDP, UNAIDS.) HIV by income data from nationally representative DHS surveys (see www.measuredhs.com). For example, HIV prevalence among women in the poorest 20 per cent of households in Cameroon is 3.1 per cent. In the richest 20 per cent it is 8.0 per cent. Côte d'Ivoire 3.6 per cent vs 8.8 per cent, Ethiopia 0.3 vs 6.1, Uganda 4.8

Chapter 4: The Naked Truth (continued)

vs 11.0, Tanzania 2.8 vs 11.4, Malawi 11 vs 18. The data for men are similar. Just 1.0 per cent of Ethiopian women with no education are HIV infected, while in women with secondary education or more, HIV is more than five times as high. Niger 0.5 per cent vs 1.2 per cent, Rwanda, 3.3 vs 6.4, Tanzania 5.8 vs 9.3.

8 De Gruttola et al., 1989.

9 De Vincenzi, 1994.

10 For two extremely comprehensive reviews of the relationship between HIV and other sexually transmitted infections, see Fleming and Wasserheit, 1999; Galvin and Cohen, 2004.

11 Bailey et al., 2007; Gray et al., 2007; Auvert et al., 2005.

12 Page-Shafer et al., 2002.

13 Buchacz et al., 2004; Tovanabutra et al., 2002; Hu et al., 2002; Bourlet et al., 2001; Quinn et al., 2000; Sabin et al., 2000; Chakraborty et al., 2001. There is an illustration which shows how viral load varies over time in the gallery section of www.wisdom ofwhores.com.

14 Brooks et al., 2006.

15 Buvé et al., 2001b; Auvert et al., 2001b; Auvert et al., 2001a; Corey et al., 2004.

16 See for example an important study investigating the differences in sexual behaviour in two high- and two low-prevalence areas in Africa. The study found that circumcision and the sexually transmitted herpes simplex virus explained more of the variation than differences in the number of sex partners (Buvé et al., 2001b).

17 Aral et al., 2005; Liljeros et al., 2003.

18 A number of international comparisons have shown that men and women in industrialized countries tend to report sex with more people over the course of their lives than men and women in Africa. Asian men (but not women) report more one-off partners – most of them prostitutes. But in the early 1990s, when HIV was just getting its teeth into Africa, men and women there were far more likely to say they had several partners on the go at once, according to a big round of comparative surveys. In Lesotho (where 23 per cent of adults are now infected with HIV), 55 per cent of men and 39 per cent of women reported more than one regular sex partner at the beginning of the 1990s, and a goodly proportion reported casual partners, too. (In these surveys, a regular partner was someone the respondent was married to, lived with, or had been sleeping with for more than a year.) In the Zambian capital Lusaka, where over a quarter of pregnant women tested positive for the virus, 22 per cent of men and half that proportion of women were doing the rounds of their regular lovers, and Tanzania looked much the same. In Côte d'Ivoire, over a third of men

Chapter 4: The Naked Truth (continued)

had more than one regular partner. And Côte d'Ivoire, with 7 per cent of all adults infected, has the highest HIV prevalence in West Africa, a region where HIV is generally rather low by the standards of the East and South of the continent. Compare this with the Asian countries in the study – the Philippines topped the charts with 3 per cent of both men and women admitting to two or more regular partners; Thailand, Sri Lanka and Singapore were all lower. Needless to say, rates of HIV infection are far, far lower in those countries too – the virus is seen in just 1.4 per cent of Thai adults, and it is virtually invisible in the other countries (Halperin and Epstein, 2004; Wellings et al., 2006; Caraël, 1995; Morris and Kretzschmar, 1997; Shelton, 2006). HIV data are from UNAIDS, 2006, with the exception of Lusaka, which is from the HIV database maintained by the US Census Bureau.

19 Wellings et al., 2006.

20 For data and references, see note 18 above and Wellings et al., 2006.

21 In the late 1990s in western Kenya, one in three nineteen-year-old girls was infected with HIV, and one in ten boys. In rural Zimbabwe, just 1 per cent of teenage boys was infected, but among girls there was seven times as much HIV. By the time they hit twenty-five, 38 per cent of girls were infected, and young men had caught up to about half the female rates (Buvé et al., 2001; Wilkinson et al., 2000). Zimbabwe data courtesy of Simon Gregson. These data appear on a graph in the gallery section of www.widomofwhores.com.

22 Gregson et al., 2002.

23 Data are from Demographic and Health Surveys (DHS) covering a nationally representative sample of adults in sixteen African countries. Examples: in Zimbabwe, 8.4 per cent of women who have never been married are infected with HIV, compared with 4.3 per cent of men who have never been married. Among those who have ever had sex, the difference is much greater: 23 per cent of sexually active single women are infected with HIV, compared with 6.2 per cent of never-married men who are not virgins. In Lesotho, 15 per cent of women who have never been married are infected with HIV, compared with 9 per cent of men. Malawi: 5 per cent vs 2 per cent, Tanzania 3.8 vs 3, Uganda 2.7 vs 0.8, Ethiopia 0.7 vs 0.3, Ghana 1.1 vs 0.3, Cameroon 3.5 vs 2.1, Côte d'Ivoire 4.6 vs 1.3, Senegal 0.3 vs zero. Only Niger has identical HIV rates in unmarried men and women, at 0.4 per cent.

24 Mishra et al., 2007. DHS data, 2003–2006. See also Pisani, 2001.

25 Carpenter et al., 1999.

26 For data and references, see note 18 above.

27 Cambodia National Center for HIV/AIDS Dermatology and STDs

Chapter 4: The Naked Truth (continued)

2004; UNAIDS, 1998; a graph showing the relationship between average number of clients and HIV prevalence can be found in the gallery section of www.wisdomofwhores.com.

28 In Australia over 62 per cent of unmarried men said in recent surveys that they used condoms the last time they had sex, and in France, close to three-quarters did. In the United States, over 70 per cent of teenage boys who had sex used condoms. Condom use by single men in several African nations has risen by between 12 and 300 per cent in recent years in countries with available data, but only three – Burkina Faso, Uganda and Zimbabwe – topped 50 per cent. Unmarried women in France and Australia are more than twice as likely to use a condom with their partners than unmarried women in South Africa. Source: Demographic and Health Survey and other national survey data, kindly analysed by Emma Slaymaker. For a full data table see www.wisdomofwhores.com/references.

29 Demographic and Health Survey data, analysed by Emma Slaymaker. Results with fewer than ten respondents not presented.

30 J. D. Shelton and B. Johnston, 2001.

31 Shafer et al., 2006; Halperin, 2006; Green et al., 2006; Stoneburner and Low-Beer, 2004a; Stoneburner and Low-Beer, 2004b; USAID, 2002; Parkhurst, 2002; Epstein, 2007.

32 Shafer et al., 2006.

33 'Sacking the wrong health minister', *The Economist*, 18 August 2007.

34 Annan, 2001.

35 See Cohen and Trussel, 1996, Chapter 4.

36 To read the text of the offending story, see www.wisdomofwhores.com/references.

37 For a fascinating history of the approach taken to infectious disease and HIV control in industrialized countries, see Baldwin, 2005.

38 HIV prevalence in Swaziland is 43 per cent. In the worst cases, where poorly stored condoms are used by couples with little experience, up to 10 per cent can break. In more normal usage, only about 2 per cent burst (see chapter 6 for references). So the chance of being exposed to HIV while using a condom in Swaziland is between 0.9 and 4.3 per cent. HIV among sex workers in China is 0.7 per cent. Infection rates in the general population are far lower (Lu et al., 2006).

39 Koop, 1988.

Chapter 5: Sacred Cows

1 Susan Hunter, *AIDS in Asia: A continent in peril*. I was quoted as saying that Indonesia has one of the fastest-growing epidemics in the world – I think the quote may have been lifted from a news report. In

Chapter 5: Sacred Cows (continued)

any case, I was obviously indulging in the kind of creative epidemiology described in Chapter 1.

2 Blower et al., 2000.

3 San Francisco Department of Public Health, 2005; Schwarcz et al., 2002; Stockman et al., 2004; McFarland and DeCarlo, 2003; Dilley et al., 2003; Kellogg et al., 1999; US CDC 1985. Measures of condom use in anal sex are not exactly comparable until 1994. Before that, condom use refers only to use with 'non-monogamous' partners. From 1994, the measure refers to unprotected anal sex with any partner over the preceding six months. From 1998, it has been possible to look at condom use by HIV status, as well as by the HIV status of partners – the same as the respondent, the opposite, or unknown. Unprotected anal sex between HIV positive men and partners who were negative or didn't know their status peaked at around 30 per cent in 2001. The rise in HIV incidence prompted renewed prevention efforts, and that 30 per cent has since fallen dramatically, to around 10 per cent in 2005. An excellent presentation of trends in behaviour and infection among gay men in San Francisco is posted under the tag 'gay' on www.wisdomofwhores.com.

4 Elford et al., 2002; Elford et al., 2001; Stolte et al., 2001; Dukers et al., 2002; Hogg et al., 2001.

5 Global HIV Prevention Working Group, 2004.

6 Martin, 2007.

7 Watts, 2003.

8 Elizabeth Rosenthal reported the political turning point in an article in the *New York Times* on 24 August 2001, under the headline: 'China Now Facing an AIDS Epidemic, a Top Aide Admits'.

9 hrw.org/english/docs/2007/05/30/global16020.htm; World Health Organization and UNAIDS, 2007.

10 See, for example, the 30 May 2007 statement from Human Rights Watch, http://hrw.org/english/docs/2007/05/30/global16020.htm.

11 Gregson et al., 2002; Vanichseni et al., 2004a; San Francisco Department of Public Health, 2005. In Surabaya in 2004/5, 50 per cent of gay guys who had had counselling and a voluntary HIV test reported unprotected anal sex in the last week, versus 37 per cent among those who hadn't tested. Among rent boys in Jakarta, 45 per cent who had tested were having risky sex, compared with 29 per cent of those who didn't know their HIV status. Among female sex workers, 53 per cent of those who had been counselled and tested for HIV voluntarily used condoms with all their clients in the last week, compared with 32 per cent of sex workers who hadn't been counselled and tested.

Chapter 5: Sacred Cows (continued)

12 The depressing slide in question can be found under the 'Evaluation' tag on www.wisdomofwhores.com.

13 Indonesia: National behavioural surveillance data for sex workers who were reached by only-government or only-NGO programmes. For sex workers in both types of programme, condom use was higher still at 83 per cent. China: Behavioural data collected by the China-UK HIV Prevention Project, selected sites in Yunnan and Sichuan, 2003.

14 See data from the Philippines Department of Health's National Epidemiology Center at: http://www2.doh.gov.ph/nec/HIV.

15 The proportion of men buying sex nationwide fell from 22 per cent to 10 per cent. Among military conscripts in the north it fell from 57 per cent to 24 per cent and among vocational students in Bangkok it fell from 31 per cent to 5 per cent. There was also a shift from buying sex from women in brothels (referred to as 'direct sex workers') to buying sex from women who worked as hostesses in clubs and karaoke bars – 'indirect sex workers'. (UNAIDS, 1998b; Phoolcharoen et al., 1998; Nelson et al., 1996; Mills et al., 1997; Hanenberg et al., 1994; Hanenberg and Rojanapithayakorn, 1998.

16 De Zoysa et al., 2005. I should note that despite this rhetorical flourish, the document later describes most of the elements of a blackmail-based programme in some detail.

17 Wolffers, 1998.

18 China-UK AIDS Prevention Project, 2003.

Chapter 6: Articles of Faith

1 George W. Bush, speaking to the Exodus Baptist Church in Philadelphia, June 2004.

2 In a nationally representative survey of over 11,400 adolescents in the United States followed in three rounds between 1995 and 2001, 21 per cent of respondents reported they had taken a virginity pledge. Some who said in early rounds that they had made such a pledge denied it in later rounds of the survey.

3 www.generationsoflight.com/generationsoflight/html/index.html. Accessed May 2004.

4 Bruckner and Bearman, 2005; Bruckner et al., 2004; Wellings et al., 2006.

5 Underhill et al., 2007.

6 Office of the United States Global AIDS Coordinator, 2004 # 8577.

7 For example, in October 2004 USAID announced the winners of the Abstinence and Healthy Choices for Youth grant, a $100 million, five-year programme run through USAID's Washington DC office. Of the eleven grant recipients announced, nine were faith-based, and

Chapter 6: Articles of Faith (continued)

all but one were US-based. www.usaid.gov/press/releases/2004/pr041005.html, accessed 22 June 2007.

8 United States, 2003. Public Law 108–25. United States Leadership Against HIV/AIDS, Tuberculosis, and Malaria Act, Washington DC. Section 402.

9 Wellings et al., 2006.

10 Cleland and Ferry, 1995; Wellings et al., 2006.

11 Ono-Kihara et al., 2002 # 7665.

12 Chen Xinxin, 'Sex and the Contemporary Chinese woman', *China Today*, 2005.

13 According to UNICEF, there were 103 girls enrolled in secondary school in Thailand through the late 1990s for every 100 boys.

14 Van Griensven et al., 2001; UNAIDS, 1998b; Thato et al., 2003. The northern study used a computerized questionnaire, so girls fessed up to a machine rather than a human being. The big leap in premarital sex probably reflects more honesty as well as more sex.

15 The numbers in detail: in 1990 57 per cent buy sex, 58 per cent of prostitutes are infected, 39 per cent of commercial sex is unprotected. The risk of exposure in commercial sex is 57% × 58% × 39% = 13%. Some 23 per cent have a girlfriend, 0.7 per cent of young women are infected, and 84 per cent of sex with girlfriends is unprotected. An additional 0.2 per cent risk of exposure.

In 1999 7 per cent buy sex, 26 per cent of prostitutes are infected, 60 per cent of commercial sex is unprotected. The risk of exposure in commercial sex is 7% × 26% × 60% = 1.1%. Some 42 per cent now have a girlfriend, 0.8 per cent of young women are infected, and 84 per cent of sex with girlfriends is unprotected. That means 0.3 per cent of young men in northern Thailand will be exposed to HIV through their girlfriends. 1990 data from Nelson et al., 1996. 1999 data from a study of vocational students in the same geographic area. Van Griensven et al., 2001.

The study populations were not identical – in 1990 the men questioned were men chosen through a random ballot of all 21-year-olds while the students questioned in 1999 were rather younger and more educated. The younger study population probably explains the lower levels in overall levels of sexual activity recorded in 1999 – just 48 per cent of the male students had had sex in 1999, compared with 87 per cent of conscripts in 1995. But even if we double overall levels of sexual activity in the student group to mirror levels of sex in the older conscript group, we get an overall risk of exposure of under three per cent. The HIV prevalence data come from the Thai national surveillance system.

Chapter 6: Articles of Faith (continued)

16 UNAIDS, 1998b. When the fall in unprotected commercial sex and the fall in STIs are set on the same graph, the parallel is striking. See the gallery section of www.wisdomofwhores.com.

17 British Broadcasting Corporation, *Panorama*, 'Sex and the Holy City', 10 October 2003.

18 Steiner et al., 1994; Steiner et al., 1993. See also www.fhi.org/en/RH/Pubs/factsheets/breakslip.htm. Condom breakage is not random – most of it is associated with improper use. In many studies, a small proportion of couples accounts for a very large proportion of reported breakage.

19 Kamenga et al., 1991; Dublin et al., 1992; Allen et al., 1992; Padian et al., 1993; Weller, 1993; Deschamps et al., 1996; Davis and Weller, 1999; Gray et al., 2001; de Vincenzi, 1994; Weller and Davis-Beaty, 2001.

20 'Stop Giving Free Condoms, Say Clerics', *The Nation* (Nairobi, Kenya), 29 November 2006. HIV data from the nationally representative Kenya Demographic and Health Survey, 2003.

21 Statement of Richard Carmona to Committee on Oversight and Government Reform, Congress of the United States, 10 July 2007. oversight.house.gov/documents/20070710111054.pdf.

22 Letter dated 18 December 2003.

23 The fact that cited data from three studies of discordant couples. In one, there were no new HIV infections in two years among 124 HIV couples who always used condoms, while in 121 couples who didn't always use condoms, 10 per cent became infected. In the second study, infection rates were 2 per cent and 12 per cent respectively, and in the third 2 per cent among consistent condom users and 14 per cent among inconsistent users. (Citations in note 19 above.) The fact sheet also cites information from thirty studies about sex education programmes. None of them increased sexual activity and those that provided detailed information about condoms and skills training both decreased sexual activity and increased condom use among those who were sexually active.

24 'Male Latex Condoms and Sexually Transmitted Diseases', CDC fact sheet, 23 January 2003: www.cdc.gov/condomeffectiveness/condoms. pdf.

25 Copies of both the old and the new fact sheets can be found at www.wisdomofwhores.com/references.

26 The last comparative data available are for 2004. In Bangladesh, just over a third of brothel-based sex workers used a condom with their most recent client, and among street-based sex workers condom use was lower. In Pakistan's two largest cities between 23 and 40 per cent of sex workers reported using condoms with their most recent client, and no client group reported more than 20 per cent use. In Indonesia,

Chapter 6: Articles of Faith (continued)

condom use with the last client had inched up to 53 per cent on average in thirteen cities by 2004.

27 Condom use at last commercial sex reported most recently by national surveillance system (data 2004 or later) was above 80 per cent in all of these countries.

28 You can find the poster in the gallery section of www.wisdomofwhores.com.

29 Weller and Davis-Beaty, 2001.

30 Sixty-seven per cent of women who asked all of their clients to use condoms used them with all clients. Less than 5 per cent of those who didn't always ask were consistent condom users.

31 Cohen, 2005.

32 www.unaids.org/en/Issues/Affected_communities, accessed April 2007.

33 For a listing of reports and documentaries on sex trafficking, see www.badasf.org/slavery/modernslavery.htm.

34 This estimate is a compilation of data from various sources, including national estimates published by the governments of Cambodia, China, Indonesia and Thailand, and research reports from India, Vietnam, Nepal and the Philippines.

35 Exactly what constitutes trafficking is not clear. For those interested in this issue, I would thoroughly recommend the works of Helen Pickering, who worked with prostitutes in both West and East Africa. She told me that when she first met a sex worker, she would often get a sob story about trickery, destitution, etc. When she got to know the girls better, she was invited home to family events and found that most sex workers came from middle-class families who were supportive of their income-generating activity. They cheerfully admitted to having made up their histories to meet the researchers' expectations. See for example Pickering et al., 1992.

36 www.etan.org/et2003/july/05-12/08claim.htm.

37 Data on earnings among factory workers in Jakarta (the highest-paying province in Indonesia), 2004, from the government statistical office, StatisticsIndonesia. Hourly 'wage' for sex work calculated from earnings reported by nearly 10,700 prostitutes in Indonesia. In 2004 clients paid an average of US$24 each. Over a quarter of these women are earning US$5.50 or less per client. Women at the bottom of the market tend to have more clients, so weekly earnings even for the lowest-paid are US$37 – more than four times what they would earn in a week in a factory.

38 Elmer, 2001; Neal et al., 2004; Monitoring the AIDS Pandemic, 2004.

39 The average hourly wage for the 73 per cent of New York City's restaurant workers who are immigrants is US$8.55, according to data

Chapter 6: Articles of Faith (continued)

from the US Census Bureau and the Economic Policy Institute. Data reproduced in Dollars and Sense: www.dollarsandsense.org/archives/2006/0906ness.html.

40 US State Department, 2006; De Zoysa et al., 2005.

41 Haugen and Hunter, 2005; Zurita, 2005.

42 The requirement came in an amendment to H.R. 1298, the United States Leadership Against HIV/AIDS, Tuberculosis, and Malaria Act. It reads: '(f) LIMITATION. – No funds made available to carry out this Act, or any amendment made to this Act, may be used to provide assistance to any group or organization that does not have a policy explicitly opposing prostitution and sex trafficking.'

43 *Asian Wall Street Journal*, 27 February 2003.

44 The head of IJM, Gary Hauger, describes one such operation in Cambodia in great detail in his book, *Terrify No More*.

45 Name changed.

46 Amy Kazmin, 'Deliver them from evil', *Financial Times*, 9 July 2004.

47 The raid is described in blow-by-bow detail by IJM President Gary Haughen in his 2005 book *Terrify No More*. When the IJM team flew in to conduct the raid, they had the foresight to bring with them a camera crew from the American TV show *Dateline NBC* (Haugen and Hunter, 2005).

48 www.gatesfoundation.org/GlobalHealth/Pri_Diseases/HIVAIDS/Announcements/Announce-060314.htm, accessed 26 May 2007.

49 The Cambodian Working Group on HIV/AIDS Projection estimates that without the aggressive condom promotion campaign, which began to reduce the rate of new infections among sex workers and their clients from the mid-1990s, there would have been nearly 1.1 million Cambodians living with HIV by the end of 2005. Largely because of vastly increased condom use in commercial sex, the true number of people infected at the end of 2005 was fewer than 120,000.

Chapter 7: HIV Shoots Up

1 Check out, for example, the 'Needle Hand-out' section of the page headed 'Harm Promotion' at www.dfaf.org/harmpromotion/maintenance.php.

2 By 2005, the most recent year for which data are available, Glasgow was giving out over a million needles a year to cover the needs of just under 5,000 injectors – around 210 per injector per year. Edinburgh estimates it has far fewer injectors – around 1,000 – but they managed to soak up around 480 needles a year each from syringe and needle programmes. Over half of these needles were given out by programmes run out of high street pharmacies. (Griesbach et al., 2006.)

Chapter 7: HIV Shoots Up (continued)

3 Davies et al., 1999; Frischer et al., 1992; Stimson, 1996; Stimson et al., 1996; Hurley et al., 1997; Buning et al., 1986; MacDonald et al., 2003; Stimson, 1995; Taylor et al., 1994; Robertson, 1990.

4 Des Jarlais et al., 2005.

5 Des Jarlais et al., 1995; Hagan et al., 1995; Vlahov et al., 1997; Buning et al., 1986; Van Den Hoek et al., 1989; Drucker et al., 1998. For a fully referenced round-up of evidence of the effect of needle exchange programmes in Asia, see Monitoring the AIDS Pandemic, 2004; Bluthenthal et al., 2000.

6 Strathdee et al., 1997. Vancouver became the first North American city to open a safe injecting room in 2003.

7 Schechter et al., 1999; Hagan et al., 1999; Bastos and Strathdee, 2000.

8 US data: Anthony et al., 1994; Canadian data: Health Canada, 1997.

9 Strathdee and Vlahov, 2001; Vietnam national sentinel surveillance data.

10 UNAIDS, 2002.

11 Sorensen and Copeland, 2000.

12 Dolan et al., 2005; Palepu et al., 2006; Stark et al., 1996; Hartel and Schoenbaum, 1998; Brugal et al., 2005; Langendam et al., 2001.

13 Wu et al, 2007.

14 In Bangkok, injectors who went to jail but said they didn't inject drugs there were twice as likely to get infected as those who didn't go to jail – the increase might be related to anal sex, though they may also have become infected in police holding cells before they went to jail.

15 McCoy et al., 1994.

16 World Health Organization et al., 2007.

17 Vanichseni et al., 2004b; Beyrer et al., 2003; Buavirat et al., 2003; Choopanya et al., 2002.

18 Dolan et al., 2003b; Dolan et al., 1996; Warren et al., 2006; Dolan et al., 2005; Heimer et al., 2006.

19 www.state.gov/s/gac/partners/guide/prevent/64035.htm.

20 www.dfaf.org/harmpromotion/maintenance.php, accessed 27 June 2007.

21 Stark et al., 2006.

22 Heimer, 1998; Brooner et al., 1998; Strathdee et al., 1999; Palepu et al., 1999; Vlahov et al., 1997; Marx et al., 2000; Doherty et al., 2000.

23 Griesbach et al., 2006; Abdulrahim et al., 2006.

24 National Commission on AIDS, 1991; Lurie and Reingold, 1993; United States General Accounting Office, 1993; United States General Accounting Office, 1998; Office of Technology Assessment of the US

Chapter 7: HIV Shoots Up (continued)

Congress, 1995; National Research Council, 1995; National Institutes of Health, 1997; American Medical Association, 1997; US Surgeon General, 2000; Institute of Medicine, 2001; Centers for Disease Control and Prevention, 2002; Institute of Medicine, 2006.

25 Data from the Indonesian national surveillance system. Related presentation on www.wisdomofwhores.com.

26 Vlahov et al., 1991; Vlahov et al., 1994; Titus et al., 1994; World Health Organization, 2004.

27 Des Jarlais et al., 2005; World Health Organization, 2006; United Nations Office on Drugs and Crime, 2007.

28 Office of the United States Global AIDS Coordinator, 2004.

29 The e-mail was leaked to public health professionals on 20 December 2004.

30 Henry Waxman wrote to US Secretary of State Condoleezza Rice on 24 June 2005 to protest at these efforts.

31 Between 1995 and 2006, the US gave UNAIDS US$225 million, around 18 per cent of its total budget. The Netherlands was next most generous at US$215 million. On a per capita basis, that in fact makes the Netherlands seventeen times more generous than the US. In 2006 the Netherlands gave the most, US$38 million, with Sweden in second place. The US ranked third at US$29.8 million, just a million dollars ahead of the UK.

32 Letter from William Steiger, Director of the US Department of Health and Human Services Office of Global Health Affairs to Denis G. Aitken, Assistant Director-General of WHO, 15 April 2004. The letter stated: 'US Government experts do not and cannot participate in WHO consultations in their individual capacity. U.S. . . . regulations require HHS [US health ministry] experts to serve as representatives of the U.S. Government at all times and advocate U.S. Government policies.'

33 Letter from Congressman Henry Waxman to Director of Health and Human Services Tommy Thompson, 28 October 2003.

34 www.ias.se/Web/WebContent/File/Old/PDF/BangkokReport.pdf.

35 Human Rights Watch, 2004.

36 World Health Organization, 2001.

37 Kral et al., 2001; Mirin et al., 1980; Mendelson and Mello, 1982.

38 In 2004/5, 54 per cent of guys who had been injecting a year or less reported multiple sex partners in the last year, compared with 50 per cent among those injecting between two and six years. By the time people had been injecting for seven or more years, 39 per cent still reported sex with more than one partner.

39 Van Ameijden et al., 1999.

Chapter 7: HIV Shoots Up (continued)

40 In a study in San Antonio, Texas, 71 per cent of regular methamphetamine users reported more than one recent sex partner, compared with just 39 per cent of heroin users (Zule and Desmond, 1999).

41 For a comprehensive summary of the data, see Monitoring the AIDS Pandemic, 2004.

42 Data from Sichuan province, courtesy of China CDC and the China UK AIDS Prevention Project. Forty seven per cent of injectors who sold unprotected sex to clients shared the needles the last time they injected, compared with just 17 per cent of female injectors who don't sell sex. Over three-quarters of sex workers who didn't inject used a condom with their most recent client, compared with just half of sex workers who inject.

43 Injectors in outpatient treatment were first tested in Jakarta in 1996/97. None of the fifty samples tested positive. The following year, still zero, out of sixty-three tests. By 2001, 105 out of 219 tests were positive (Pisani et al., 2003c).

Chapter 8: Ants in the Sugar-Bowl

1 Spending for 2007 was estimated on the basis of pledges and commitments already made by donors, as well as from budgetary allocations and projections made by developing country governments (UNAIDS, 2005).

2 United Nations Office on Drugs and Crime, 2006.

3 Ghana figures compiled by David Wilson, Global HIV/AIDS Programme, World Bank.

4 www.pepfar.gov/countries/c19637.htm.

5 Nigeria Demographic and Health Survey, 2003. In addition, 22 per cent of men and just 2 per cent of women contacted in households said they had multiple partners in the last year. Men always report more partners than women, but that is a huge discrepancy by any standards. Polygamy may account for part of the difference, but it is likely that some of the women that guys reported as girlfriends were actually prostitutes. Levels of reported condom use would suggest so – men reported almost the same level of condom use with their last non-marital partner (46 per cent) as they did with a sex worker (49 per cent). If even half the men reporting multiple partners sometimes paid for sex, it would mean 4 million clients of sex workers.

6 US Census Bureau HIV Database. The Saki study was among just sixty-one women. In Ibadan, 243 prostitutes were tested.

7 Adelekan and Lawal, 2006.

8 These are my calculations, using data from the OECD's development assistance committee database. The database is incomplete because

Chapter 8: Ants in the Sugar-Bowl (continued)

donors do not always report their spending accurately. I therefore report trends over time (which are less affected by under-reporting) rather than dollar amounts. See Kaiser Family Foundation, 2007 for a discussion of the limitations of the database.

9 Incidence estimates are made using the model described in Pisani et al., 2003b. Sources for finance data as follows: China: supplied by Salil Panakadan of the UNAIDS office in Beijing. Thailand: Tangcharoensathien et al., 2006. Cambodia: Martin, 2005.

10 Huffam et al., 2002.

11 Quoted in a UNDP press release, 3 July 2002.

12 George W. Bush, speaking on 31 May 2007.

13 Pisani et al., 2006.

14 If you're interested in the issues, see www.wisdomofwhores.com/references.

15 Recently, USAID has started buying some condoms in Asia because American manufacturers can't keep up with demand. See Celia Dugger, 'U.S. Jobs Shape Condoms' Role in Foreign Aid', *New York Times*, 26 October 2006.

16 Even the US government's auditors complain about how clumsy and wasteful these procurement procedures are. See United States General Accounting Office, 2001.

17 Médecins Sans Frontières, 2003.

18 Information on current purchases of generics from the PEPFAR website, www.pepfar.gov/press/83466.htm, accessed 1 June 2007.

19 World Health Organization, 2007.

20 Global Fund site accessed 23 June 2007. George Bush speech 31 May 2007.

21 Avahan monitoring data, September 2007.

22 www.theglobalfund.org/programs/grantdetails.aspx?compid=375&grantid=139&lang=en&CountryId=IND, accessed 2 June 2007. The Global Fund is absolutely exemplary in terms of the information it makes available to the public. All grant applications, evaluations and 'scorecards' are posted on its well-organized website.

23 My notes from a World Bank STI programme review debriefing, 10 February 1999.

24 For examples of the sort of analysis that will lose you friends, see www.wisdomofwhores.com/references.

Chapter 9: Full Circle

1 The *Kompas* piece appeared on 29 September 2001. It took two years of nagging before that statement was translated into official policy. The National Narcotics Control Board signed a memorandum of

Chapter 9: Full Circle (continued)

understanding with the National AIDS Commission formalizing harm reduction as a national strategy for drug users on 8 December 2003.

2 Examples from Indonesia: WHO paid the Population Council to adapt and translate guidelines for behavioural surveillance produced by WHO's Southeast Asian Regional Office, even though Indonesia already had its own, far more advanced guidelines, produced by the Bureau of Statistics together with the Ministry of Health and paid for by USAID. UNAIDS organized a series of meetings and hired a consultant to develop an Indonesian version of something called the Country Response Information System, a database pushed by Geneva, even though Indonesia's Bureau of Statistics was already developing a more appropriate and comprehensive database in consultation with the Ministry of Health and all major organizations working in AIDS, including UNAIDS.

3 E-mail, Chris Green to EP, 17 December 2003.

4 At the time, there were no systematic estimates of the number of people in high-risk groups. There were also wildly differing views about the number of people likely to have been infected through the dodgy blood transfusion practices. Chinese colleagues claimed the number was 200,000 or fewer, while some foreign consultants guesstimated it as at least 900,000, and possibly much higher. Mass screenings carried out in 2004 suggest the lower figure was much closer to the truth.

5 UN Theme Group on HIV/AIDS in China, 2001.

6 Wu et al., 2007.

BIBLIOGRAPHY

Abdulrahim, D., D. Gordon and D. Best (2006). Findings of a survey of needle exchanges in England. London, National Treatment Agency for Substance Misuse.

Adelekan, M. L. and R. A. Lawal (2006). 'Drug use and HIV infection in Nigeria. A review of recent findings.' *African Journal of Drug & Alcohol Studies* 5(2).

Adu-Oppong, A., R. M. Grimes, M. W. Ross, et al. (2007). 'Social and behavioral determinants of consistent condom use among female commercial sex workers in Ghana.' *AIDS Educ Prev* 19(2): 160–72.

Allen, S., J. Tice, P. Van de Perre, et al. (1992). 'Effect of serotesting with counselling on condom use and seroconversion among HIV discordant couples in Africa.' *BMJ* 304(6842): 1605–9.

American Medical Association (1997). Report 8 of the Council on Scientific Affairs (A-97). Chicago, American Medical Association.

Annan, K. (2001). Address made by Secretary-General Kofi Annan to the African Summit on HIV/AIDS, Tuberculosis and Other Infectious Diseases in Abuja, Nigeria. Abuja.

Anthony, J., L. Warner and R. Kessler (1994). 'Comparative Epidemiology of Dependence on Tobacco, Alcohol, Controlled Substances and Inhalants.' *Experimental and Clinical Psychopharmacology* 3(3): 244–68.

Aral, S. O., N. S. Padian and K. K. Holmes (2005). 'Advances in multilevel approaches to understanding the epidemiology and prevention of sexually transmitted infections and HIV: an overview.' *J Infect Dis* 191 Suppl 1: S1–6.

Attaran, A. and J. Sachs (2001). 'Defining and refining international donor support for combating the AIDS pandemic.' *Lancet* 357(9249): 57–61.

Auvert, B., A. Buve, B. Ferry, et al. (2001a). 'Ecological and individual level analysis of risk factors for HIV infection in four urban populations in sub-Saharan Africa with different levels of HIV infection.' *AIDS* 15 Suppl 4: S15–30.

Auvert, B., A. Buvé, E. Lagarde, et al. (2001b). 'Male circumcision and HIV infection in four cities in sub-Saharan Africa.' *Aids* 15 Suppl 4: S31-40.

Auvert, B., D. Taljaard, E. Lagarde, et al. (2005). 'Randomized, controlled intervention trial of male circumcision for reduction of HIV infection risk: the ANRS 1265 Trial.' *PLoS Med* 2(11): e298.

Bailey, R. C., S. Moses, C. B. Parker, et al. (2007). 'Male circumcision for HIV prevention in young men in Kisumu, Kenya: a randomised controlled trial.' *Lancet* 369(9562): 643–56.

Baldwin, P. (2005). *Disease and Democracy. The Industrialized world faces AIDS*. Berkley CA, University of California Press.

Barnett, T. and A. Whiteside (2006). *AIDS in the 21st Century: Disease and Globalisation, 2nd Edition*. London, Palgrave Macmillan.

Bastos, F. I. and S. A. Strathdee (2000). 'Evaluating effectiveness of syringe exchange programmes: current issues and future prospects.' *Soc Sci Med* 51(12): 1771–82.

Beyrer, C., J. Jittiwutikarn, W. Teokul, et al. (2003). 'Drug use, increasing incarceration rates, and prison-associated HIV risks in Thailand.' *AIDS Behav* 7(2): 153–61.

Blacker, J. and B. Zaba (1997). 'HIV prevalence and the life-time risk of dying from AIDS.' *Health Transition Review* 7(Suppl 2).

Blower, S. M., H. B. Gershengorn and R. M. Grant (2000). 'A tale of two futures: HIV and antiretroviral therapy in San Francisco.' *Science* 287(5453): 650–54.

Bluthenthal, R. N., A. H. Kral, L. Gee, et al. (2000). 'The effect of syringe exchange use on high-risk injection drug users: a cohort study.' *AIDS* 14(5): 605–11.

Bourlet, T., C. Cazorla, P. Berthelot, et al. (2001). 'Compartmentalization of HIV-1 according to antiretroviral therapy: viral loads are correlated in blood and semen but poorly in blood and saliva.' *AIDS* 15(2): 284–5.

Brooks, J. T., K. E. Robbins, A. S. Youngpairoj, et al. (2006). 'Molecular analysis of HIV strains from a cluster of worker infections in the adult film industry, Los Angeles 2004.' *AIDS* 20(6): 923–8.

Brooner, R., M. Kidorf, V. King, et al. (1998). 'Drug abuse treatment success among needle exchange participants.' *Public Health Rep* 113 Suppl 1: 129–39.

Brown, P. (1992). 'World AIDS Programme "Lacks Vision".' *New Scientist* 1806: 14.

Bruckner, H. and P. Bearman (2005). 'After the promise: the STD consequences of adolescent virginity pledges.' *J Adolesc Health* 36(4): 271–8.

Bruckner, H., A. Martin and P. S. Bearman (2004). 'Ambivalence and pregnancy: adolescents' attitudes, contraceptive use and pregnancy.' *Perspect Sex Reprod Health* 36(6): 248–57.

Brugal, M. T., A. Domingo-Salvany, R. Puig, et al. (2005). 'Evaluating the impact of methadone maintenance programmes on mortality due to overdose and AIDS in a cohort of heroin users in Spain.' *Addiction* 100(7): 981–9.

Buavirat, A., K. Page-Shafer, G. J. van Griensven, et al. (2003). 'Risk of prevalent HIV infection associated with incarceration among injecting drug users in Bangkok, Thailand: case-control study.' *BMJ* 326(7384): 308.

Buchacz, K., P. Patel, M. Taylor, et al. (2004). 'Syphilis increases HIV viral load and decreases CD4 cell counts in HIV-infected patients with new syphilis infections.' *AIDS* 18(15): 2075–9.

Buning, E. C., R. A. Coutinho, G. H. van Brussel, et al. (1986). 'Preventing AIDS in drug addicts in Amsterdam.' *Lancet* 1(8495): 1435.

Burton, A. H. and T. E. Mertens (1998). 'Provisional country estimates of prevalent adult human immunodeficiency virus infections as of end 1994: a description of the methods.' *Int J Epidemiol* 27(1): 101–7.

Buvé, A., M. Caraël, R. J. Hayes, et al. (2001). 'Multicentre study on factors determining differences in rate of spread of HIV in sub-Saharan Africa: methods and prevalence of HIV infection.' *AIDS* 15 Suppl 4: S5–14.

Buvé, A., M. Caraël, R. J. Hayes, et al. (2006). 'Multicentre study on factors determining differences in rate of spread of HIV in sub-Saharan Africa: summary and conclusions.' *AIDS* 15 Suppl 4: S127–31.

Cambodia National Center for HIV/AIDS Dermatology and STDs, (2004). BSS V: Sexual Behaviour among Urban Sentinel Groups, Cambodia 2001, Phnom Penh.

Caraël, M. (1995). 'Sexual behaviour', in *Sexual Behaviour and AIDS in the Developing World*. J. Cleland and B. Ferry (eds.). London, Taylor & Francis: 75–123.

Carpenter, L. M., A. Kamali, A. Ruberantwari, et al. (1999). 'Rates of HIV-1 transmission within marriage in rural Uganda in relation to the HIV sero-status of the partners.' *AIDS* 13(9): 1083–9.

Centers for Disease Control and Prevention (2002). Syringe Exchange Programs. Atlanta, CDC.

Chakraborty, H., P. K. Sen, R. W. Helms, et al. (2001). 'Viral burden in genital secretions determines male-to-female sexual transmission of HIV-1: a probabilistic empiric model.' *AIDS* 15(5): 621–7.

Chin, J. (2007). *The AIDS Pandemic: The Collision of Epidemiology with Political Correctness*. Oxford, Radcliffe Publishing.

Chin, J. and J. Mann (1989). 'Global surveillance and forecasting of AIDS.' *Bulletin Of The World Health Organization* 67(1): 1–7.

China–UK AIDS Prevention Project (2003). Situational Analysis of Sexual Health in Sichuan and Yunnan. Beijing, China UK.

Choopanya, K., D. C. Des Jarlais, S. Vanichseni, et al. (2002). 'Incarceration and risk for HIV infection among injection drug users in Bangkok.' *J Acquir Immune Defic Syndr* 29(1): 86–94.

Civic, D. and D. Wilson (1996). 'Dry sex in Zimbabwe and implications for condom use.' *Soc Sci Med* 42(1): 91–8.

Cleland, J. and B. Ferry (1995). *Sexual Behavior and AIDS in the Developing World (Complete book)*. London, Taylor & Francis.

Cohen, B. and J. Trussell (1996). Chapter 4: Sexual behavior and HIV/AIDS. *Preventing and Mitigating AIDS in Sub-Saharan Africa: Research and Data Priorities for the Social and Behavioral Sciences*. G. Cohen and J. Trussell. Washington, D.C., National Academy Press: 105–154.

Cohen, S. (2005). 'Ominous Convergence: Sex Trafficking, Prostitution and International Family Planning.' *The Guttmacher Report on Public Policy* (February 2005).

Corey, L., A. Wald, C. L. Celum, et al. (2004). 'The effects of herpes simplex virus-2 on HIV-1 acquisition and transmission: a review of two overlapping epidemics.' *J Acquir Immune Defic Syndr* 35(5): 435–45.

Davies, A. G., R. M. Cormack and A. M. Richardson (1999). 'Estimation of injecting drug users in the City of Edinburgh, Scotland, and number infected with human immunodeficiency virus.' *Int J Epidemiol* 28(1): 117–21.

Davis, K. R. and S. C. Weller (1999). 'The effectiveness of condoms in reducing heterosexual transmission of HIV.' *Fam Plann Perspect* 31(6): 272–9.

De Gruttola, V., G. R. d. Seage, K. H. Mayer, et al. (1989). 'Infectiousness of HIV between male homosexual partners.' *J Clin Epidemiol* 42(9): 849–56.

De Vincenzi, I. (1994). 'A longitudinal study of human immunodeficiency virus transmission by heterosexual partners. European Study Group on Heterosexual Transmission of HIV.' *N Engl J Med* 331(6): 341–6.

De Zoysa, I., W. Rojanapithayakorn, J. Smarajit, et al. (2005). Review of the 100% Targeted Condom Promotion Programme in Myanmar. Yangon, World Health Organization.

Des Jarlais, D. C., H. Hagan, S. R. Friedman, et al. (1995). 'Maintaining low HIV seroprevalence in populations of injecting drug users.' *JAMA* 274(15): 1226–31.

Des Jarlais, D. C., T. Perlis, K. Arasteh, et al. (2005). 'HIV incidence among injection drug users in New York City, 1990 to 2002: use of serologic test algorithm to assess expansion of HIV prevention services.' *Am J Public Health* 95(8): 1439–44.

Deschamps, M. M., J. W. Pape, A. Hafner, et al. (1996). 'Heterosexual transmission of HIV in Haiti.' *Ann Intern Med* 125(4): 324–30.

Dilley, J. W., W. J. Woods, J. Sabatino, et al. (2003). 'Availability of combination therapy for HIV: effects on sexual risk taking in a sample of high-risk gay and bisexual men.' *AIDS Care* 15(1): 27–37.

Doherty, M. C., B. Junge, P. Rathouz, et al. (2000). 'The effect of a needle exchange program on numbers of discarded needles: a 2-year follow-up.' *Am J Public Health* 90(6): 936–9.

Dolan, K., W. Hall and A. Wodak (1996). 'Methadone maintenance reduces injecting in prison.' *BMJ* 312(7039): 1162.

Dolan, K., S. Rutter and A. D. Wodak (2003a). 'Prison-based syringe exchange programmes: a review of international research and development.' *Addiction* 98(2): 153–8.

Dolan, K. A., J. Shearer, M. MacDonald, et al. (2003b). 'A randomised controlled trial of methadone maintenance treatment versus wait list control in an Australian prison system.' *Drug Alcohol Depend* 72(1): 59–65.

Dolan, K. A., J. Shearer, B. White, et al. (2005). 'Four-year follow-up of imprisoned male heroin users and methadone treatment: mortality, re-incarceration and hepatitis C infection.' *Addiction* 100(6): 820–28.

Doll, R. and A. B. Hill (1950). 'Smoking and carcinoma of the lung; preliminary report.' *BMJ* 2(4682): 739–48.

Drucker, E., P. Lurie, A. Wodak, et al. (1998). 'Measuring harm reduction: the effects of needle and syringe exchange programs and methadone maintenance on the ecology of HIV.' *AIDS* 12(Suppl A): S217–30.

Dublin, S., P. S. Rosenberg and J. J. Goedert (1992). 'Patterns and predictors of high-risk sexual behavior in female partners of HIV-infected men with hemophilia.' *AIDS* 6(5): 475–82.

Dukers, N. H., J. Spaargaren, R. B. Geskus, et al. (2002). 'HIV incidence on the increase among homosexual men attending an Amsterdam sexually transmitted disease clinic: using a novel approach for detecting recent infections.' *AIDS* 16(10): F19–24.

Elford, J., G. Bolding and L. Sherr (2002). 'High-risk sexual behaviour increases among London gay men between 1998 and 2001: what is the role of HIV optimism?' *AIDS* 16(11): 1537–44.

Elford, J., S. Leaity, F. Lampe, et al. (2001). 'Incidence of HIV infection among gay men in a London HIV testing clinic, 1997–1998.' *AIDS* 15(5): 650–53.

Elmendorf, E., E. R. Jensen and E. Pisani (2004). Evaluation of World Bank Assistance In Responding to the AIDS Epidemic: Indonesia Case Study. Washington DC, World Bank.

Elmer, L. (2001). HIV/AIDS Intervention Data on Commercial Sex Workers in Vietnam: A Review of Recent Research Findings: ii–v, 1–20.

Epstein, H. (2007). *The Invisible Cure: AIDS in Africa*. New York, Farrar, Straus & Giroux.

Family Health International (2000). Behavioral surveillance surveys (BSS): guidelines for repeated behavioral surveys in populations at risk of HIV. Arlington VA.

Family Health International (2002). Estimating the Size of Populations at Risk for HIV: Issues and Methods, A joint UNAIDS/IMPACT/FHI workshop: Report and Conclusions. Bangkok.

Family Health International (2006). First things first: Guidelines on management and coding of behavioural surveillance data. Bangkok.

Fleming, D. T. and J. N. Wasserheit (1999). 'From epidemiological synergy to public health policy and practice: the contribution of other sexually transmitted diseases to sexual transmission of HIV infection.' *Sexually Transmitted Infections* 75(3): 3–17.

Food and Agriculture Organization (2005). Impact of HIV/AIDS on fishing communities: Policies to Support Livelihoods, Rural Development and Public Health. Rome.

Frischer, M., S. T. Green, D. J. Goldberg, et al. (1992). 'Estimates of HIV infection among injecting drug users in Glasgow, 1985–1990.' *AIDS* 6(11): 1371–5.

Galvin, S. R. and M. S. Cohen (2004). 'The role of sexually transmitted diseases in HIV transmission.' *Nat Rev Microbiol* 2(1): 33–42.

Glantz, S. A., John Slade, Lisa A. Bero, et al. (eds.) (1996). *The Cigarette Papers*. Berkeley CA, University of California Press.

Global HIV Prevention Working Group (2004). HIV Prevention in an Era of Expanded Treatment. Seattle WA.

Government of Swaziland (2006). The Second National Multisectoral HIV and AIDS Strategic Plan 2006–2008. Mbabane, National Emergency Response Council on HIV and AIDS.

Gray, R. H., G. Kigozi, D. Serwadda, et al. (2007). 'Male circumcision for HIV prevention in men in Rakai, Uganda: a randomised trial.' *The Lancet* 369(9562): 657–66.

Gray, R. H., M. J. Wawer, R. Brookmeyer, et al. (2001). 'Probability of HIV-1 transmission per coital act in monogamous, heterosexual, HIV-1-discordant couples in Rakai, Uganda.' *The Lancet* 357(9263): 1149–53.

Green, E. C., D. T. Halperin, V. Nantulya, et al. (2006). 'Uganda's HIV prevention success: the role of sexual behavior change and the national response.' *AIDS Behav* 10(4): 335–46; discussion 347–50.

Gregson, S., C. A. Nyamukapa, G. P. Garnett, et al. (2002). 'Sexual mixing patterns and sex-differentials in teenage exposure to HIV infection in rural Zimbabwe.' *The Lancet* 359(9321): 1896–1903.

Griesbach, D., D. Abdulrahim, D. Gordon, et al. (2006). Needle Exchange

Provision in Scotland: A Report of the National Needle Exchange Survey. Edinburgh, Scottish Executive.

Hagan, H., D. C. Des Jarlais, S. R. Friedman, et al. (1995). 'Reduced risk of hepatitis B and hepatitis C among injection drug users in the Tacoma syringe exchange program.' *Am J Public Health* 85(11): 1531–7.

Hagan, H., J. P. McGough, H. Thiede, et al. (1999). 'Syringe exchange and risk of infection with hepatitis B and C viruses.' *Am J Epidemiol* 149(3): 203–13.

Halperin, D. T. (1999). 'Dry sex practices and HIV infection in the Dominican Republic and Haiti.' *Sex Transm Infect* 75(6): 445–6.

Halperin, D. T. (2006). 'The controversy over fear arousal in AIDS prevention and lessons from Uganda.' *J Health Commun* 11(3): 266–7.

Halperin, D. T. and H. Epstein (2004). 'Concurrent sexual partnerships help to explain Africa's high HIV prevalence: implications for prevention.' *Lancet* 364(9428): 4–6.

Hamilton, W. T. and D. Kessler (2004). 'BMJ papers could include honesty box for research warts.' *BMJ* 328(1320): 7451.

Hanenberg, R. and W. Rojanapithayakorn (1998). 'Changes in prostitution and the AIDS epidemic in Thailand.' *AIDS Care* 10(1): 69–79.

Hanenberg, R. S., W. Rojanapithayakorn, P. Kunasol, et al. (1994). 'Impact of Thailand's HIV-control programme as indicated by the decline of sexually transmitted diseases.' *The Lancet* 344(8917): 243–5.

Hartel, D. M. and E. E. Schoenbaum (1998). 'Methadone treatment protects against HIV infection: two decades of experience in the Bronx, New York City.' *Public Health Rep* 113 Suppl 1: 107–15.

Haugen, G. and G. Hunter (2005). *Terrify No More*. Nashville TN, W Publishing Group.

Health Canada (1997). Canada's Alcohol and Other Drug Survey, 1994. Ottawa.

Heimer, R. (1998). 'Can syringe exchange serve as a conduit to substance abuse treatment?' *J Subst Abuse Treat* 15(3): 183–91.

Heimer, R., H. Catania, R. G. Newman, et al. (2006). 'Methadone maintenance in prison: evaluation of a pilot program in Puerto Rico.' *Drug Alcohol Depend* 83(2): 122–9.

Heng, S., B. Hor and P. Gorbach (2001). Cambodian household male survey (BSS IV 2000). Phnom Penh, National Centre for HIV/AIDS, Dermatology and STDs (NCHADS).

Hogg, R. S., A. E. Weber, K. Chan, et al. (2001). 'Increasing incidence of HIV infections among young gay and bisexual men in Vancouver.' *AIDS* 15(10): 1321–2.

Hu, D. J., S. Subbarao, S. Vanichseni, et al. (2002). 'Higher viral loads and other risk factors associated with HIV-1 seroconversion during a period of high incidence among injection drug users in Bangkok.' *J Acquir Immune Defic Syndr* 30(2): 240–7.

Huffam, S., B. J. Currie, P. Knibbs, et al. (2002). 'HIV-1 infection in foreign nationals working in East Timor.' *Lancet* 360(9330): 416.

Hull, T., E. Sulistyaningsih and G. Jones (1997). *Pelacuran di Indonesia*. Jakarta, Sinar Harapan.

Human Rights Watch (2004). Not enough graves: The war on drugs, HIV/AIDS and the violation of Human Rights. New York.

Hunter, S. (2003). *Black Death: AIDS in Africa*. London, Palgrave Macmillan.

Hurley, S. F., D. J. Jolley and J. M. Kaldor (1997). 'Effectiveness of needle exchange programmes for prevention of HIV infection.' *The Lancet* 349(9068): 1797–1800.

Indonesia Directorate General of Communicable Disease Control and Environmental Health (2003). National Estimates of Adult HIV Infection, Indonesia 2002. Jakarta, Ministry of Health.

Institute of Medicine (2001). No Time to Lose: Getting More from HIV Prevention. Washington DC, National Academy of Sciences.

Institute of Medicine (2006). Preventing HIV Infection among Injecting Drug Users in High Risk Countries: An Assessment of the Evidence. Washington DC, National Academy of Sciences.

Jalal, F., H. M. Abednego, T. Sadjimin, et al. (1994). 'HIV and AIDS in Indonesia.' *AIDS* 8(Suppl 2): S91–4.

Kaiser Family Foundation, UNAIDS. (2007). Financing the response to AIDS in low- and middle-income countries: International assistance from the G8, European Commission and other donor Governments, 2006. Geneva.

Kamenga, M., R. W. Ryder, M. Jingu, et al. (1991). 'Evidence of marked sexual behavior change associated with low HIV-1 seroconversion in 149 married couples with discordant HIV-1 serostatus: experience at an HIV counselling center in Zaire.' *AIDS* 5(1): 61–7.

Kellogg, T. A., W. McFarland and M. H. Katz (1999). 'Recent increases in incidence of seroconversion among repeat anonymous testers in San Francisco.' *AIDS* 13(16): 2302–4.

Koop, C. (1988). *Individual freedom and the public interest*. First International Conference on the Global Impact of AIDS, London, 8–10 March 1988, Alan R. Liss.

Kral, A. H., R. N. Bluthenthal, J. Lorvick, et al. (2001). 'Sexual transmission of HIV-1 among injection drug users in San Francisco, USA: risk-factor analysis.' *The Lancet* 357(9266): 1397–1401.

Kumalawati, J., N. Sukartini and E. Donegan (2002). Report on HIV test kit Evaluation for Government of Indonesia. Jakarta, World Health Organization, Family Health International.

Langendam, M. W., G. H. van Brussel, R. A. Coutinho, et al. (2001). 'The impact of harm-reduction-based methadone treatment on mortality among heroin users.' *Am J Public Health* 91(5): 774–80.

Liljeros, F., C. R. Edling, H. E. Stanley, et al. (2003). 'Social networks: sexual contacts and epidemic thresholds.' *Nature* 423(6940): 606.

Linnan, M. (1992). AIDS in Indonesia: The Coming Storm. Jakarta, USAID.

Lu, F., N. Wang, Z. Wu, et al. (2006). 'Estimating the number of people at risk for and living with HIV in China in 2005: methods and results.' *Sex Transm Infect* 82 Suppl 3: iii, 87–91.

Lubis, I., J. Master, A. Munif, et al. (1997). 'Second report of AIDS related attitudes and sexual practices of the Jakarta WARIA (male transvestites) in 1995.' *Southeast Asian J Trop Med Public Health* 28(3): 525–9.

Lurie, P. and A. Reingold (eds.) (1993). *The public health impact of needle exchange programs in the United States and abroad, Vol. 1.* Atlanta, Centers for Disease Control and Prevention.

MacDonald, M., M. Law, J. Kaldor, et al. (2003). 'Effectiveness of needle and syringe programmes for preventing HIV transmission.' *International Journal of Drug Policy* 14(5–6): 353–7.

Martin, G. (2007). AIDS Expenditure in the Asia Region — some preliminary findings. Draft report, Background paper for the Comission on AIDS in Asia.

Martin, G. H. (2005). Technical Note: Tracking HIV/AIDS Resources in Cambodia. Washington DC, Policy Project, Futures Group.

Marx, M. A., B. Crape, R. S. Brookmeyer, et al. (2000). 'Trends in crime and the introduction of a needle exchange program.' *Am J Public Health* 90(12): 1933–6.

McCoy, C. B., J. E. Rivers, H. V. McCoy, et al. (1994). 'Compliance to bleach disinfection protocols among injecting drug users in Miami.' *J Acquir Immune Defic Syndr* 7(7): 773–6.

McFarland, W. and P. DeCarlo. (2003). 'What is the effect of HIV treatment on HIV prevention?' http://www.caps.ucsf.edu/pubs/FS/treatment.php.

Médecins Sans Frontières (2003). Untangling the web of price reductions: a pricing guide for the purchase of ARVs for developing countries. Geneva, Médecins Sans Frontières, Campaign for Access to Essential Medicines.

Mendelson, J. H. and N. K. Mello (1982). 'Hormones and psycho-sexual development in young men following chronic heroin use.' *Neurobehav Toxicol Teratol* 4(4): 441–5.

Merson, M. H. (2006). 'The HIV-AIDS pandemic at 25 – the global response.' *N Engl J Med* 354(23): 2414–7.

Mills, S., P. Benjarattanaporn, A. Bennett, et al. (1997). 'HIV risk behavioral surveillance in Bangkok, Thailand: sexual behavior trends among eight population groups.' *AIDS* 11 Suppl 1: S43–51.

Mills, S., T. Saidel, A. Bennett, et al. (1998). 'HIV risk behavioral surveillance: a methodology for monitoring behavioral trends.' *AIDS* 12 Suppl 2: S37–46.

Mirin, S. M., R. E. Meyer, J. H. Mendelson, et al. (1980). 'Opiate use and sexual function.' *Am J Psychiatry* 137(8): 909–15.

Mishra, V., M. Vaessen, S. Bignami-Van Assche, et al. (2007). 'Why Do So Many HIV-Discordant Couples in Sub-Saharan Africa Have Female Partners Infected, Not Male Partners?' *Submitted.*

Monitoring the AIDS Pandemic (2004). AIDS in Asia: Face the Facts. Bangkok.

Monitoring the AIDS Pandemic Network (MAP) (2001). The status and trends of the HIV/AIDS epidemics in Asia and the Pacific. Melbourne, Australia: 40 pages.

Morris, M. and M. Kretzschmar (1997). 'Concurrent partnerships and the spread of HIV.' *Aids* 11(5): 641–8.

National Centre in HIV Epidemiology and Clinical Research (2006). HIV/AIDS, viral hepatitis and sexually transmissible infections in Australia, Annual Surveillance Report 2006. Sydney, University of New South Wales.

National Commission on AIDS (1991). The Twin Epidemics of Substance Use and HIV. Washington DC.

National Institutes of Health (1997). Interventions to prevent HIV risk behaviours. Bethesda MD.

National Intelligence Council (2000). The Global Infectious Disease Threat and Its Implications for the United States. Langley VA.

National Research Council (1995). Needle Exchange Programs Reduce HIV Transmission among People Who Inject Illegal Drugs. Washington DC.

Neal, J. J., H. Sopheab, V. Chhum, et al. (2004). *Integrated HIV, STD, and Behavioral Survey among Female Sentinel Groups, Banteay Meanchey Province, Cambodia, 2003*. Abstract submitted to 15th International AIDS Conference, Bangkok.

Nelson, K. E., D. D. Celentano, S. Eiumtrakol, et al. (1996). 'Changes in sexual behavior and a decline in HIV infection among young men in Thailand [see comments].' *N Engl J Med* 335(5): 297–303.

Office of Technology Assessment of the US Congress (1995). The Effectiveness of AIDS Prevention Efforts. Washington DC, US Government Printing Office.

Office of the United States Global AIDS Coordinator (2004). The President's Emergency Plan for AIDS Relief. U.S. Five-Year Global HIV-AIDS Strategy. Washington DC.

Okunlola, M. A., I. O. Morhason-Bello, K. M. Owonikoko, et al. (2006). 'Female condom awareness, use and concerns among Nigerian female undergraduates.' *J Obstet Gynaecol* 26(4): 353–6.

Ono-Kihara, M., M. Kihara and H. Yamazaki (2002). 'Sexual Practices and the Risk for HIV/STDs Infection of Youth in Japan.' *JMAJ* 45(12): 520–25.

Padian, N. S., T. R. O'Brien, Y. Chang, et al. (1993). 'Prevention of heterosexual transmission of human immunodeficiency virus through couple counseling.' *J Acquir Immune Defic Syndr* 6(9): 1043–8.

Page-Shafer, K., C. H. Shiboski, D. H. Osmond, et al. (2002). 'Risk of HIV infection attributable to oral sex among men who have sex with men and in the population of men who have sex with men.' *AIDS* 16(17): 2350–52.

Palepu, A., S. A. Strathdee, R. S. Hogg, et al. (1999). 'The social determinants of emergency department and hospital use by injection drug users in Canada.' *J Urban Health* 76(4): 409–18.

Palepu, A., M. W. Tyndall, R. Joy, et al. (2006). 'Antiretroviral adherence and HIV treatment outcomes among HIV/HCV co-infected injection drug users: the role of methadone maintenance therapy.' *Drug Alcohol Depend* 84(2): 188–94.

Parkhurst, J. O. (2002). 'The Ugandan success story? Evidence and claims of HIV-1 prevention.' *The Lancet* 360(9326): 78–80.

Phoolcharoen, W., K. Ungchusak, W. Sittitrai, et al. (1998). 'Thailand: Lessons from a strong national response to HIV/AIDS.' *AIDS* 12 (Suppl B): S123–35.

Pickering, H., J. Todd, D. Dunn, et al. (1992). 'Prostitutes and their clients: a Gambian survey.' *Soc Sci Med* 34(1): 75–88.

Pisani, E. (2001). HIV in the 21st Century: Implications for UNICEF. New York, UNICEF.

Pisani, E., Dadun, P. K. Sucahya, et al. (2003a). 'Sexual behavior among injection drug users in 3 Indonesian cities carries a high potential for HIV spread to noninjectors.' *J Acquir Immune Defic Syndr* 34(4): 403–6.

Pisani, E., G. P. Garnett, N. C. Grassly, et al. (2003b). 'Back to basics in HIV prevention: focus on exposure.' *BMJ* 326(7403): 1384–7.

Pisani, E., P. Girault, M. Gultom, et al. (2004). 'HIV, syphilis infection, and sexual practices among transgenders, male sex workers, and other men who have sex with men in Jakarta, Indonesia.' *Sex Transm Infect* 80(6): 536–40.

Pisani, E., H. Purnomo, A. Sutrisna, et al. (2006). 'Basing policy on evidence: low HIV, STIs and risk behaviour in Dili, East Timor argue for more focused interventions.' *Sex Transm Infect* 82: 88–93.

Pisani, E., A. Setiawan, O. Harun, et al. (2003c). *Sex or drugs? Hobson's choice for harm reduction programmes*. 14th International Conference on the Reduction of Drug-related Harm, Chiang Mai.

Quinn, T. C., M. J. Wawer, N. Sewankambo, et al. (2000). 'Viral load and heterosexual transmission of human immunodeficiency virus type 1. Rakai Project Study Group.' *N Engl J Med* 342(13): 921–9.

Robertson, R. (1990). 'The Edinburgh epidemic: a case study', in *AIDS and Drug Misuse: The Challenge for Policy and Practice in the 1990s*. J. Strang and G. Stimson (eds.). New York, Routledge: 95–107.

Sabin, C. A., H. Devereux, A. N. Phillips, et al. (2000). 'Course of viral load throughout HIV-1 infection.' *J Acquir Immune Defic Syndr* 23(2): 172–7.

San Francisco Department of Public Health (2005). HIV/AIDS Epidemiology Annual Report 2005. San Francisco, HIV/AIDS Statistics, Epidemiology, and Intervention Research Section.

Sandala, L., P. Lurie, M. R. Sunkutu, et al. (1995). '"Dry sex" and HIV infection among women attending a sexually transmitted diseases clinic in Lusaka, Zambia.' *AIDS* 9 Suppl 1: S61–8.

Schechter, M. T., S. A. Strathdee, P. G. Cornelisse, et al. (1999). 'Do needle exchange programmes increase the spread of HIV among injection drug users? An investigation of the Vancouver outbreak.' *AIDS* 13(6): F45–51.

Schwarcz, S. K., T. A. Kellogg, W. McFarland, et al. (2002). 'Characterization of sexually transmitted disease clinic patients with recent human immunodeficiency virus infection.' *J Infect Dis* 186(7): 1019–22.

Schwartländer, B., P. D. Ghys, E. Pisani, et al. (2001a). 'HIV surveillance in hard-to-reach populations.' *AIDS* 15 Suppl 3: S1–3.

Schwartländer, B., K. A. Stanecki, T. Brown, et al. (1999). 'Country-specific estimates and models of HIV and AIDS: methods and limitations.' *AIDS* 13(17): 2445–58.

Schwartländer, B., J. Stover, N. Walker, et al. (2001b). 'AIDS. Resource needs for HIV/AIDS.' *Science* 292(5526): 2434–6.

Shafer, L. A., S. Biraro, A. Kamali, et al. (2006). *HIV prevalence and incidence are no longer falling in Uganda – A case for renewed prevention efforts. Evidence from a rural population cohort 1989–2005 and from ANC surveillance*. Int Conf AIDS, Toronto.

Shelton, J. D. (2006). 'Confessions of a condom lover.' *The Lancet* 368(9551): 1947–9.

Sorensen, J. L. and A. L. Copeland (2000). 'Drug abuse treatment as an HIV prevention strategy: a review.' *Drug Alcohol Depend* 59(1): 17–31.

South Africa Department of Health (2006). National HIV Survey, 2005. Pretoria.

Stark, K., U. Herrmann, S. Ehrhardt, et al. (2006). 'A syringe exchange programme in prison as prevention strategy against HIV infection and hepatitis B and C in Berlin, Germany.' *Epidemiol Infect* 134(4): 814–19.

Stark, K., R. Muller, U. Bienzle, et al. (1996). 'Methadone maintenance treatment and HIV risk-taking behaviour among injecting drug users in Berlin.' *J Epidemiol Community Health* 50(5): 534–7.

Steiner, M., C. Piedrahita, C. Joanis, et al. (1993). 'Can Condom Users Likely to Experience Condom Failure Be Identified?' *Family Planning Perspectives* 25(5): 220–3, 226.

Steiner, M., C. Piedrahita, C. Joanis, et al. (1994). 'Condom Breakage and Slippage Rates Among Study Participants in Eight Countries.' *International Family Planning Perspectives* 20(2): 55–8.

Stimson, G. V. (1995). 'AIDS and injecting drug use in the United Kingdom, 1987–1993: the policy response and the prevention of the epidemic.' *Soc Sci Med* 41(5): 699–716.

Stimson, G. V. (1996). 'Has the United Kingdom averted an epidemic of HIV-1 infection among drug injectors?' *Addiction* 91(8): 1085–8; discussion 1089–99.

Stimson, G. V., G. M. Hunter, M. C. Donoghoe, et al. (1996). 'HIV-1 prevalence in community-wide samples of injecting drug users in London, 1990–1993.' *AIDS* 10(6): 657–66.

Stockman, J., S. K. Schwarcz, L. M. Butler, et al. (2004). 'HIV prevention fatigue among high-risk populations in San Francisco.' *J Acquir Immune Defic Syndr* 35(4): 432–3.

Stolte, I. G., N. H. Dukers, J. B. de Wit, et al. (2001). 'Increase in sexually transmitted infections among homosexual men in Amsterdam in relation to HAART.' *Sex Transm Infect* 77(3): 184–6.

Stoneburner, R. L. and D. Low-Beer (2004a). 'Population-level HIV declines and behavioral risk avoidance in Uganda.' *Science* 304(5671): 714–18.

Stoneburner, R. L. and D. Low-Beer (2004b). 'Sexual partner reductions explain human immunodeficiency virus declines in Uganda: comparative analyses of HIV and behavioural data in Uganda, Kenya, Malawi, and Zambia.' *Int J Epidemiol* 33(3): 624.

Strathdee, S. A., D. D. Celentano, N. Shah, et al. (1999). 'Needle-exchange attendance and health care utilization promote entry into detoxification.' *J Urban Health* 76(4): 448–60.

Strathdee, S. A., D. M. Patrick, S. L. Currie, et al. (1997). 'Needle exchange is not enough: lessons from the Vancouver injecting drug use study.' *AIDS* 11(8): F59–65.

Strathdee, S. A. and D. Vlahov (2001). 'The effectiveness of needle exchange programs: A review of the science and policy.' *AIDScience* 1(16).

Synergy and USAID (2001). The United States Responds to the Global AIDS Pandemic. Washington DC, USAID.

Tangcharoensathien, V., H. Chokchaicharn, K. Tisayaticom, et al. (2006). Asia Pacific Regional Report on National AIDS Spending Assessment 2000–2004. Bangkok, International Health Policy Program, and Thailand National Economic and Social Development Board.

Taylor, A., M. Frischer, S. T. Green, et al. (1994). 'Low and stable prevalence of HIV among drug injectors in Glasgow.' *Int J STD AIDS* 5(2): 105–7.

Telles Dias, P. R., K. Souto and K. Page-Shafer (2006). 'Long-term female condom use among vulnerable populations in Brazil.' *AIDS Behav* 10(4 Suppl): S67–75.

Thato, S., D. Charron-Prochownik, L. D. Dorn, et al. (2003). 'Predictors of condom use among adolescent Thai vocational students.' *J Nurs Scholarsh* 35(2): 157–63.

Thomsen, S. C., W. Ombidi, C. Toroitich-Ruto, et al. (2006). 'A prospective study assessing the effects of introducing the female condom in a sex worker population in Mombasa, Kenya.' *Sex Transm Infect* 82(5): 397–402.

Titus, S., M. Marmor, D. Des Jarlais, et al. (1994). 'Bleach use and HIV seroconversion among New York City injection drug users.' *J Acquir Immune Defic Syndr* 7(7): 700–704.

Tovanabutra, S., V. Robison, J. Wongtrakul, et al. (2002). 'Male viral load and heterosexual transmission of HIV-1 subtype E in northern Thailand.' *J Acquir Immune Defic Syndr* 29(3): 275–83.

UN Theme Group on HIV/AIDS in China (2001). HIV/AIDS: China's Titanic Peril 2001: Update of the AIDS Situation and Needs Assessment Report, UNAIDS China Office.

UNAIDS (1998a). AIDS Epidemic Update. December 1998. Geneva.

UNAIDS (1998b). 'Relationship of HIV and STD declines in Thailand to behavior change – A synthesis of existing studies.' *UNAIDS Key Materials Publication*, Geneva.

UNAIDS (1998c). 'Report of the global HIV/AIDS epidemic.' *UNAIDS/WHO Publication*, Geneva.

UNAIDS (2002). Preliminary Overview Market Prices of Opioid Agonist Treatment Drugs. Draft. Vienna.

UNAIDS (2003). Estimating HIV infection in a concentrated epidemic: Lessons from Indonesia. Geneva.

UNAIDS (2005). Resource needs for an expanded response to AIDS in low- and middle-income countries. Geneva.

UNAIDS (2006). Report on the global AIDS epidemic 2006. Geneva.

UNAIDS and WHO (1996). The HIV/AIDS Situation in mid-1996. Global and Regional highlights. Geneva, UNAIDS.

United Nations Office on Drugs and Crime (2006). World Drug Report 2006. Vienna, UNODC.

United Nations Office on Drugs and Crime (2007). 'Epidemiology of drug use in Iran.' Vienna.

Underhill, K., P. Montgomery and D. Operario (2007). 'Sexual abstinence only programmes to prevent HIV infection in high income countries: systematic review.' *BMJ* 335(7613): 248.

United States (2003). Public Law 108–25. United States Leadership Against HIV/AIDS, Tuberculosis, and Malaria Act, Washington DC.

United States (2003). United States Leadership Against HIV/AIDS, Tuberculosis, and Malaria Act. 22 *USC 7601*.

United States General Accounting Office (1993). Needle Exchange Programs: Research Suggests Promise as an AIDS Prevention Strategy. Washington DC.

United States General Accounting Office (1998). Needle Exchange Programs in America: Review of Published Studies and On-going Research. *Report to the Committee on Appropriations for the Departments of Labor, Health and Human Services, Education and Related Agencies*. Washington DC, US General Accounting Office.

United States General Accounting Office (2001). U.S. Agency for International Development Fights AIDS in Africa, but Better Data Needed to Measure Impact. Washington DC.

United States Centers for Disease Control and Prevention (1985). 'Epidemiologic Notes and Reports Self-Reported Behavioral Change Among Gay and Bisexual Men – San Francisco.' *MMWR* 34(40): 613–15.

United States State Department (2006). Trafficking in Persons Report, June 2006. Washington DC.

United States Surgeon General (1964). Smoking and Health: Report of the Advisory Committee to the Surgeon General of the Public Health Service. Washington DC.

United States Surgeon General (2000). Evidence-based findings on the efficacy of syringe exchange programs: an analysis of the scientific research completed since April 1998. Washington, DC, United States Department of Health & Human Services.

USAID (2002). What happened in Uganda? Declining HIV Prevalence, Behavior Change, and the National Response. Washington DC, USAID.

Van Ameijden, E. J., M. W. Langendam, J. Notenboom, et al. (1999). 'Continuing injecting risk behaviour: results from the Amsterdam Cohort Study of drug users.' *Addiction* 94(7): 1051–61.

Van Den Hoek, J., H. Van Haastrecht and R. Coutinho (1989). 'Risk Reduction Among Intravenous Drug Users in Amsterdam Under the

Influence of AIDS.' *American Journal of Public Health* 79(10): 1355–7.

Van Griensven, F., S. Supawitkul, P. H. Kilmarx, et al. (2001). 'Rapid assessment of sexual behavior, drug use, human immunodeficiency virus, and sexually transmitted diseases in northern Thai youth using audio-computer-assisted self-interviewing and noninvasive specimen collection.' *Pediatrics* 108(1): E13.

Vanichseni, S., D. C. Des Jarlais, K. Choopanya, et al. (2004a). 'Sexual Risk Reduction in a Cohort of Injecting Drug Users in Bangkok, Thailand.' *J Acquir Immune Defic Syndr* 37(1): 1170–79.

Vanichseni, S., F. van Griensven, B. Metch, et al. (2004b). 'Risk Factors and Incidence of HIV Infection Among Injecting Drug Users (IDUs) Participating in the AIDSVAX B/E HIV Vaccine Efficacy Trial, Bangkok.' 15th International AIDS Conference, Bangkok.

Vlahov, D., J. Astemborski, L. Solomon, et al. (1994). 'Field effectiveness of needle disinfection among injecting drug users.' *J Acquir Immune Defic Syndr* 7(7): 760–66.

Vlahov, D., B. Junge, R. Brookmeyer, et al. (1997). 'Reductions in high-risk drug use behaviors among participants in the Baltimore needle exchange program.' *J Acquir Immune Defic Syndr Hum Retrovirol* 16(5): 400–406.

Vlahov, D., A. Munoz, D. D. Celentano, et al. (1991). 'HIV seroconversion and disinfection of injection equipment among intravenous drug users, Baltimore, Maryland.' *Epidemiology* 2(6): 444–6.

Warren, E., R. Viney, J. Shearer, et al. (2006). 'Value for money in drug treatment: economic evaluation of prison methadone.' *Drug Alcohol Depend* 84(2): 160–66.

Watts, J. (2003). 'China faces up to HIV/AIDS epidemic. World AIDS day is marked by launch of huge public-awareness campaign.' *The Lancet* 362(9400): 1983.

Welbourn, A. (2006). 'Sex, Life and the Female Condom: Some Views of HIV Positive Women.' *Reprod Health Matters* 14(28): 32–40.

Weller, S. and K. Davis-Beaty (2001). 'Condom effectiveness in reducing heterosexual HIV transmission.' *The Cochrane Database of Systematic Reviews* 2007(2).

Weller, S. C. (1993). 'A meta-analysis of condom effectiveness in reducing sexually transmitted HIV.' *Soc Sci Med* 36(12): 1635–44.

Wellings, K., M. Collumbien, E. Slaymaker, et al. (2006). 'Sexual behaviour in context: a global perspective.' *Lancet* 368(9548): 1706–28.

Wilkinson, D., S. S. Abdool Karim, B. Williams, et al. (2000). 'High HIV incidence and prevalence among young women in rural South Africa: developing a cohort for intervention trials.' *J Acquir Immune Defic Syndr* 23(5): 405–9.

Witte, S. S., N. El-Bassel, L. Gilbert, et al. (2006). 'Promoting female condom use to heterosexual couples: findings from a randomized clinical trial.' *Perspect Sex Reprod Health* 38(3): 148–54.

Wolffers, I. (1998). 'Why this initiative is important.' *Research for Sex Work*: 1–16.

World Bank (1996). Staff appraisal report: Indonesia HIV/AIDS and STDs prevention and management project. Washington DC.

World Bank (1997). *Confronting AIDS: Public priorities in a global epidemic*. Oxford, Oxford University Press.

World Health Organization (2001). HIV/AIDS in Asia and the Pacific Region, World Health Organization, Regional Offices for the Western Pacific and for South-East Asia.

World Health Organization (2004). Effectiveness of Sterile Needle and Syringe Programming in Reducing HIV/AIDS among Injecting Drug Users. Geneva.

World Health Organization (2006). A Best Practice Review of HIV Prevention and Care among Injecting Drug Users in the Islamic Republic of Iran. Cairo, WHO Regional Office for the Eastern Mediterranean.

World Health Organization (2007). Towards Universal Access: Scaling up priority HIV/AIDS interventions in the health sector Progress Report. Geneva.

World Health Organization and UNAIDS (2007). Guidance on provider-initiated HIV testing and counselling in health facilities. Geneva.

World Health Organization, UNAIDS and UNODC (2007). Effectiveness of Interventions to Manage HIV in Prisons – Needle and syringe programmes and bleach and decontamination strategies. Geneva.

Wu, Z., S. G. Sullivan, Y. Wang, et al. (2007). 'Evolution of China's response to HIV/AIDS.' *The Lancet* 369(9562): 679–90.

Zhang, D., P. Bi, F. Lv, et al. (2007). 'Changes in HIV prevalence and sexual behavior among men who have sex with men in a northern Chinese city: 2002–2006.' *J Infect*.

Zule, W. A. and D. P. Desmond (1999). 'An ethnographic comparison of HIV risk behaviors among heroin and methamphetamine injectors.' *Am J Drug Alcohol Abuse* 25(1): 1–23.

Zurita, B. (2005, 14/12). 'Prostitution is Not a Profession.' http:/www.cwfa.org/articles/9691/BLI/family/index.htm. Retrieved 27/05/07, 2007.

INDEX

abstinence, 188–9, 198
Acquired Immune Deficiency Syndrome *see* HIV/AIDS
Acting for Women in Distressing Situations (AFESIP), 223, 224, 225
Africa
  condom use, 143, 144, 146, 148, 206
  HIV/AIDS, 12, 19–20, 124–8, 134, 136–8, 142, 143–53, 156–8, 159–60, 271, 317
  sexual behaviour, 134–8, 142, 150
Ahnaf, Arizal, 304
AIDS *see* HIV/AIDS
Akers HIV test kits, 113–15
Altman, Lawrence, 31
Alzety (Al), 98–100, 105, 180, 317, 324
American Medical Association, 251
Arifin, Luwi, 194
Arifin, Renjani, 194
anal sex, 58–9, 130, 153
Annan, Kofi, 149, 152, 160, 260, 295
antiretroviral treatment, 162–8, 285–7, 312–13
a'Nzeki, Ndingi Mwana, 202
Asia
  condom use, 206
  gay scene, 74–5
  HIV/AIDS, 12, 138–40, 272–3
  sexual behaviour 193–199
Asia Pacific Network of Sex Workers, 167
Asian Epidemic Model, 306–7
Asian Harm Reduction Network, 229, 230–1
Association of People With AIDS in Kenya (TAPWAK), 186

Baasyir, Abu Bakar, 207
Ball, Andrew, 29–30
Balthazar, Godfrey, 110
banci see waria
Bangladesh
  condom use, 206
  HIV/AIDS, 127, 296–7
Bardon, Jeanine, 278, 291, 300
Beckham, David, 124
'Beltway Bandits', 276, 290, 291
Bennett, Bill, 205
Bertini, Adriana, 258
Bickerton, Ashley, 194
Bill and Melinda Gates Foundation, 226, 270, 294–5, 323
Bono, 124, 308
Botswana: HIV/AIDS, 19, 127
Brazil
  antiretroviral drugs, 285–6
  HIV/AIDS, 19, 23, 221
breastfeeding, 131
Britain
  HIV/AIDS, 8–9
  overseas aid, 284, 293
  prisons, 245–6

British Medical Journal, 85
Brown, Leticia, 66
Brown, Tim, 306–7
Buffett, Warren, 294
'Bullshit Bingo', 296
Burma: HIV/AIDS, 19, 29, 76
Burton, Sir Richard, 150
Bush, George W., 33–4, 188–9, 191–2, 203, 285, 289, 308
'Buy American Act', 283–4

Cambodia
  condom use, 221–2
  HIV/AIDS, 139, 140, 185, 273
  'rescuing' of prostitutes, 223–6
Cameroon: HIV/AIDS, 128, 143
Caraël, Michel, 37
Carletonville, South Africa, 137
Carmona, Richard, 203
case-control studies, 3–4, 5–6
Catholic Relief Services, 288
CDC *see* Centers for Disease Control and Prevention
Centers for Disease Control and Prevention, United States (CDC), 6, 204, 251, 256
Charles, Robert, 255
Chiang Mai, Thailand, 177, 196, 197
China
  drug injection, 267
  HIV/AIDS, 139, 156–7, 168–71, 176, 273, 318–24
  methadone use, 241
  needle exchange programmes, 232
  sex industry, 180–1
  sexual behaviour, 195–6
Church World Services (CWS), 120
Cipinang prison, Jakarta, 244, 247
circumcision, 130, 313
Club Volvo, Hong Kong, 67–8
cocaine, 76, 234–5, 249
cohort studies, 4, 5, 234
commercial sex
  'abolition' of, 219–20
  'loyalty oath', 220–1
  remuneration for, 216–19
condom use
  Africa, 143, 144, 146, 148, 206
  Asia, 206
  Cambodia, 221–2
  in commercial sex, 200
  cost-benefit analysis, 209–13
  opposition to, 200–8
  promotion, 311–12
  Indonesia, 65–6, 200, 203, 206–7, 210–11
  San Francisco, 164–5
  Thailand, 199–200
Confronting AIDS (World Bank), 125
Costa, Antonio Maria, 255
Cox, Adrian, 232–5, 265, 317
CWS *see* Church World Services

Dali, China, 288–9
dangdut (Indonesian pop music), 63
data management and analysis, 106–109
de Araujo, Rui Maria, 275–6
de Vincenzi, Isabelle, 29, 130
Department for International Development (DfID), UK, 289, 291
Department of Social Affairs, Jakarta, 53, 55
Desi, 78, 92, 105, 244, 317
Dialma, Emmanuel, 223, 225
Dili, East Timor, 215–16, 278–80
discordant couples, 138
DKT, 221
Dlamini, Gugu, 155
Dolly, Surabaya (red light district), 101
Donegan, Elizabeth, 114
Dongxing, China, 180–1, 217–18
Drug Free America Foundation, 230, 248
drug injection
  drug warriors vs harm reductionists, 228–31, 234, 250
  and HIV/AIDS, 30, 75–6, 154, 304–5

needle exchange programmes, 231–9, 248, 249–54, 268, 312
patterns of, 235–6
and sex, 263–8

East Timor
HIV/AIDS, 139, 278–81
HIV/AIDS prevention programme, 274–6, 281–3, 287–8
sex trafficking, 215–16
Eastern Europe: HIV/AIDS, 12, 25, 26, 32
The Economist, 152–3, 320
Ecstasy, 76
Eli Lilly, 285
epidemiology, 1
principles of, 3–7
Epidemiology (journal), 7
Eris, 91, 116
estimates of HIV/AIDS, 21–26, 35, 318–24
estimates of at-risk populations 94–97

Fahdli, 252–253, 269
Family Health International (FHI), 41, 44, 48, 57, 65, 99, 174–5, 206, 277–8, 291, 302
FHI *see* Family Health International
Fifteenth International AIDS Conference, Bangkok (2004), 258–61
Fox, David, 2, 15, 40–1, 89–90, 262–3
Food and Agriculture Organization, 270–1
FPI *see* Islamic Defenders Front
Frankie, 243–244
Fuad, 50–51, 59–60, 72–73, 266
Futures Group, The, 291

Gay Men's Health Crisis, 8, 174
Gay Related Immune Deficiency Syndrome (GRIDS), 8
Geldof, Bob, 124
General Accounting Office, United States, 251
Ghana: HIV/AIDS, 271
Gilead Sciences, 224
GIPA *see* Greater Involvement of People with AIDS
Girault, Philippe, 69–70, 89–90, 278
Global Fund for AIDS, TB and Malaria, 261, 271, 273, 289, 291, 295–8, 308–9
Global Programme on AIDS, World Health Organisation, 13, 14, 154
Golden Lotus, 195–6
gonorrhoea, 164
Gordon, David, 79, 80
Graham, Billy, 34
Graham, Franklin, 34
Granat (anti-drugs group), 182
Grand cinema, Jakarta, 70–1
Greater Involvement of People with AIDS (GIPA), 183
Green, Chris, 236, 305
GRIDS *see* Gay Related Immune Deficiency Syndrome
Guangxi province, China, 232
Guinea: HIV/AIDS, 127

Hansen, Keith, 305–6
Hardjo, Happy, 304
harm reduction, 228–39, 234, 248–54, 268, 312
Harmayn, Shanty, 88
Haugen, Garry, 219
Helms, Jesse, 34
Henan province, China, 170
heroin, 76, 77–83, 234–5, 249, 263–6
herpes, 130, 133
HIV/AIDS
causes, 9, 311, 316–7
as development problem, 125–7, 144, 157–60, 209, 275, 316
estimates, 21–26, 35, 318–24
facts vs myth, 128–34

HIV/AIDS – *continued*
first recorded cases, 8–9, 153
and inequalities of wealth and gender, 125–7, 144, 157–60, 209, 275, 316
and inter-agency competition, 289–92
prevention programmes, 18–19, 309–16, 324–5
as 'punishment', 153, 155–6
resource allocation, 33–4, 191–2, 269–70, 272–3, 274–5, 295–6, 298–9, 308–9
surveillance, 37–9, 41–2, 49, 84–7, 89
testing for, 35–6, 109–18, 132, 168–73
Ho Chi Minh City: needle exchange programme, 237
homosexuality, 8, 58–9, 69–75, 98–100, 154, 173–4
Hong Kong, 1–2, 67–8, 78, 318
HSV *see* herpes
Huang, Stephen, 74
Hudner, Tim, 215–16
Hudson, Rock, 8
Hughes, Donna, 221–2, 227
Human Immunodeficiency Virus *see* HIV/AIDS

IDU *see* drug injection
IEC (information, education and communication), 57
IJM *see* International Justice Mission
India: HIV/AIDS, 28, 140, 295
Indonesia
blood testing, 111–15, 116–18
condom use, 65–6, 200, 203, 206–7, 210–11
female sex workers, 60, 64–5, 173, 266
gay scene, 73–5
heroin use, 77–83, 263–6
HIV prevention programmes, 45–6, 98–100, 248, 296, 302–3
HIV surveillance programmes, 41–2, 43–4, 46, 48–50, 85–6, 100–5, 107–9, 118–19, 120–3, 304–5
law-and-order crackdowns, 90–1
male sex workers, 69–72
National Anti-AIDS Movement, 200
needle exchange programmes, 252
prisons, 122, 241–8
Red Cross, 111, 112, 115
role of NGOs, 175, 176
sex industry, 44–5, 60–6, 69–72, 95–7
transgender sex workers, *see* waria
waria, 46–8, 50–60, 87–90, 168
Indonesian Mujahiddin Council (MMI), 207
Ines Angela, 87–88, 317
Institute of Medicine, United States 251
International AIDS Society, 259
International AIDS Vaccine Initiative, 309
International Community of Women Living with HIV/AIDS, 151–2
International Justice Mission (IJM), 219, 222, 225–6
Iran: needle exchange programme, 253–4
Islam: opposition to condom use, 200–1, 207
Islamic Defenders Front (FPI), 91
Iswandono, Naning, 304
Italian Cultural Centre, Jakarta, 87, 189
Ivory Coast: HIV/AIDS, 138

Jamieson, Nancy, 278
Japan: sexual behaviour, 194–5

Kadarisman, Paul, 194
Kaleeba, Chris, 145, 149
Kaleeba, Noerine, 136, 149–50
Kalijodo, Jakarta (red light district), 120

Kenya: HIV/AIDS, 26, 110–11, 138, 148, 186–7, 202
Kerobokan prison, Indonesia, 243
ketamine, 76
Kilapong, Kiki, 113, 200
Kompas (Indonesian newspaper), 302
Koop, Everett, 157
Kramat Tunggak, Jakarta (red light district), 62

Laksmindra, Nungky, 79, 194
The Lancet, 7
Langherhans' cells, 130
Latin America: HIV/AIDS, 12, 19, 25, 26, 249
'Lazarus transformation', 167–8
Li, Ronald, 68
Ling Ling, 238–240, 301, 317
Linnan, Mike, 45
London School of Hygiene and Tropical Medicine, 3, 9
'loyalty oath', 220–1
Lu, Fan, 321–2
Lukman, 64, 119–20

Madlala-Routledge, Noziwe, 148
Malawi: HIV/AIDS, 19
Mann, Jonathan, 154
Masaka district, Uganda, 147
Mbeki, Thabo, 148, 317
Meiling, 218–219
Merauke, Papua, 101–5
methadone, 239–41, 247, 312
methamphetamines, 76
Mills, Steve, 41
MMI *see* Indonesian Mujahiddin Council
Monasch, Roeland, 29
Moonlight (Jakarta disco), 116, 117
Morning in America (US radio show), 205
Mozambique: HIV/AIDS, 148
MSM *see* homosexuality *and* waria
Muhammad, Marie, 112
Murphy, Dan, 282–3
Museveni, Yoweri, 147, 317

Nana, 65–66, 87, 92, 158, 210–11, 245, 301, 324
Nancy, 52–56, 59, 88
National Academy of Sciences, United States, 251, 256
National Bureau of Statistics, Indonesia, 100
National Commission on AIDS, United States, 251
National Institutes of Health, United States, 251, 256–7
National Intelligence Council, United States, (NIC), 32–3
National Network of People Living with HIV and AIDS (Swaziland), 151
needle exchange programmes, 231–9, 248–54, 268, 312
Neilsen, Graham, 17
Nepal: HIV/AIDS, 19, 76
New England Journal of Medicine, 7
New York: needle exchange programmes, 231–2
New York Times, 31, 32, 169
New York Yankees, 229
NGOs *see* non-governmental organizations
NIC *see* National Intelligence Council
Nigeria: HIV/AIDS, 24–5, 271–2
non-governmental organizations (NGOs): role in HIV/AIDS prevention, 174–80, 287–92

Office of Technology Assessment of the US Congress, 251
Open Society Institute, 221
oral sex, 131
Otto, Brad, 122
Over, Mead, 125–7, 142

Pakistan
  condom use, 206
  HIV/AIDS, 140
Palfrey, Deborah, 220

Panorama (BBC TV programme), 201
Papua, 101–5
Partnership for a Drug Free America, 229
patient activism, 183–7
peer education, 180–3
Pelangi, 98
Pen, Mony, 185
People Living With HIV/AIDS (PWHA), 292
PEPFAR *see* President's Emergency Plan for AIDS Relief
philanthropy, 294–5, 308
Philippines: HIV/AIDS, 139, 140, 176
Piot, Peter, 14, 18, 24, 30, 31, 255, 256
Pisani, Roger 229–30, 250
PMTCT *see* Prevention of Mother To Child Transmission
population control, 2
Population Services International, 291
porn stars, 133
Prabawanti, Ciptasari, 48
pregnancy, 131, 313
Prego (Italian bar in Jakarta), 74, 98, 99, 100
premarital sex, 193
 and HIV/AIDS, 196–9
President's Emergency Plan for AIDS Relief (PEPFAR), 192, 205, 220, 272, 281, 285–6, 289, 290
Prevention of Mother To Child Transmission (PMTCT), 239
prevention programmes for HIV, 18–19, 309–10
Prie, Mitu, 194
prisons: prevalence of HIV/AIDS in, 122, 241–8
prostitution *see* commercial sex
public health, 7, 154–5
Purdy, Chris, 200, 207
PWHA *see* People Living With HIV/AIDS

qualitative research, 49

racism: fear of in AIDS industry, 152–3
Rahman, Lisabona (Lisa), 194
Rawa Bebek, Jakarta (red light district), 120
Rawa Malang, Jakarta (red light district), 62–6, 119–20, 245
Reagan, Ronald, 8, 18, 149
Rees, Grover Joseph, 281–3
religious attitudes to HIV/AIDS, 188, 191–2
resource allocation, 33–4, 191–2, 269–70, 272–3, 274–5, 295–6, 298–9, 308–9
Roman Catholic Church: opposition to condom use, 201–3
Rosenthal, Elizabeth, 169
Russia: HIV/AIDS, 26, 76, 245

Sachs, Jeffrey, 33
Saidel, Tobi, 41
San Francisco: gay scene, 164–5, 173
SARS, 154
Schwartländer, Bernhard, 23–4
Scotland: needle exchange programmes, 231
Senegal: HIV/AIDS, 31, 144–6, 156, 324
Setiawan, Made, 48
sex trafficking, 213–16, 223–4
sexual behaviour
 abstinence, 188–9, 198
 Africa, 134–8, 142, 150
 China, 195–6
 and drug injection, 263–8
 Japan, 194–5
 surveys, 91–2, 94
sexually transmitted infections (STIs), 18, 44, 45, 57–8, 122, 133–4, 141–2, 205, 280, 307, 317
Shelley, John, 246
Shinawatra, Thaksin, 259–60, 261
Sierra Leone: HIV/AIDS, 127

Silfanus, Fonny, 107
Sin, Cardinal, 207
Smaradyota, Arya, 74
Smith, Chris, 281–2
smoking, 4–6
Soepanto, Arwati, 48
Somalia: HIV/AIDS, 127
Sonagachi, Calcutta, 177
South Africa: HIV/AIDS, 29, 124–5, 127, 148
Spiritia Foundation, 305
Standing Committee on Nutrition, United Nations, 270
Stanecki, Karen, 260–1
STIs *see* sexually transmitted infections
Stover, John, 306–7
sub-Saharan Africa: HIV/AIDS, 19, 33, 42, 143, 149
Sugiharto, Lenny, 52–3, 55–8, 90, 166–8, 198
Suharto, President, 44
Suharto, Tommy, 244, 247
Suharto, Tutut, 112
Sukarnoputri, Megawati, 116
Supartono, Alex, 194
Surabaya, Indonesia, 101, 175, 181–2, 252–3, 263–5
surveillance programmes for HIV, 37–9, 39–42, 49, 84–7, 89
Sutiyoso (governor of Jakarta), 62
Suwannawong, Paisan, 260–1
Suwastoyo, Bhimanto, 73–4, 98
Swaziland: HIV/AIDS, 125, 148, 151–2, 155, 156
syphilis, 35, 57–8, 130, 154, 164, 284

T cells, 130
Talenta, Surabaya (drop-in centre for drug users), 181–2
Tanamur disco, Jakarta, 73–4
TAPWAK *see* Association of People With AIDS in Kenya
TASO *see* The AIDS Support Organization
Thailand
  100 pct condom policy, 178–9
  condom use, 178–9, 199–200
  drugs policy, 259–61
  HIV/AIDS, 19, 139, 140, 166, 173, 178–9, 196–7, 273
  sex-change operations, 54
  sex industry, 176–9

Thatcher, Margaret, 159, 253–4, 293
The AIDS Support Organization (TASO), 149
'Three Cs boyfriends', 136–7
'Three Ones' principle, 289
Tobias, Randall, 220
Traditional Values Coalition, 257
Transgender sex workers *see* waria
trantib ('discipline police'), 90
Trujillo, Alfonso Lopez, 201
Tshabalala-Msimang, Manto, 148
tuberculosis, 247

Uganda: HIV/AIDS, 8, 19, 36, 138, 144–7, 156, 315
Ukraine: HIV/AIDS, 19, 76, 245
UNAIDS
  author's reports for, 34–5
  'Best Practice' reports, 119
  campaigns, 20–2
  establishment of, 13, 14–16
  estimates of HIV infection, 22–6
  Global Reports, 26, 28, 30–1
  jargon, 16–18
  Reference Group on HIV Estimates, Modelling and Projections, 125
  on sex trafficking, 213
  and US AIDS policy, 255
UNESCO, 13
UNICEF, 13, 288
United Nations Development Programme, 13, 275
United Nations Office on Drugs and Crime (UNODC), 78, 255

United Nations Population Fund, 13, 166
United States Surgeon General, 4, 6, 251
US Congress, 33
US Food and Drug Administration, 284
USA
  HIV/AIDS, 8, 19, 153, 154, 204
  policy on AIDS prevention, 254–8
  spending on AIDS prevention, 192, 270, 273–5, 283–6
USAID, 33

Vancouver: needle exchange programmes, 232, 234–5, 237
Village Voice, 8
viral load, 131–2, 163
virginity pledges, 190
voluntary confidential testing, 169, 172–3

Walker, Neff, 23–4, 319
waria, 46–8, 50–60, 87–90, 168
Waxman, Henry, 203–4, 255
Wellcome Trust, 309
WHO *see* World Health Organization
Wignall, Steve, 113–14
William J Clinton Foundation, 270
Wolfensohn, James, 32
Wolffers, Ivan, 180
Women's Network for Unity (Cambodia), 224
World AIDS Day, 21
World Bank, 31–2, 45, 108, 110, 125–7, 305
World Health Organization (WHO), 13–14, 16, 107, 166, 172, 179, 256, 288
Wu, Zunyou, 323

Yunnan province, China, 237–8, 288–9

Zambia: HIV/AIDS, 19, 129
Zimbabwe: HIV/AIDS, 19, 148, 173
Zurita, Brenda, 219